A NETWORKED SELF AND HUMAN AUGMENTICS, ARTIFICIAL INTELLIGENCE, SENTIENCE

Every new technology invites its own sets of hopes and fears, and raises as many questions as it answers revolving around the same theme: Will technology fundamentally alter the essence of what it means to be human? This volume draws inspiration from the work of the many luminaries who approach augmented, alternative forms of intelligence and consciousness. Scholars contribute their thoughts on how human augmentic technologies and artificial or sentient forms of intelligence can be used to enable, reimagine, and reorganize how we understand our selves, how we conceive the meaning of "human", and how we define meaning in our lives.

Zizi Papacharissi is Professor and Head of the Communication Department and Professor of Political Science at the University of Illinois-Chicago, and University Scholar at the University of Illinois System. Her work focuses on the social and political consequences of online media. She has published nine books, including *Affective Publics, A Private Sphere, A Networked Self: Identity, Community, and Culture on Social Network Sites* (Routledge, 2010) and over 60 journal articles, book chapters and reviews. She is the founding and current editor of the open access journal *Social Media and Society*.

A Networked Self

Each volume in this series develops and pursues a distinct theme focused on the concept of the Networked Self. The five volumes cover the broad range of socio-cultural, political, economic, and sociotechnical issues that shape and are shaped by the (networked) self in late modernity—what we have come to describe as the anthropocene.

A Networked Self: Identity, Community and Culture on Social Network Sites
A Networked Self and Platforms, Stories, Connections
A Networked Self and Love
A Networked Self and Birth, Life, Death
A Networked Self and Human Augmentics, Artificial Intelligence, Sentience

Growing upon the initial volume, *A Networked Self: Identity, Community and Culture on Social Network Sites*, published in 2010, the five volumes will form a picture of the way digital media shape contemporary notions of identity.

A NETWORKED SELF AND HUMAN AUGMENTICS, ARTIFICIAL INTELLIGENCE, SENTIENCE

Edited by
Zizi Papacharissi

Routledge
Taylor & Francis Group

NEW YORK AND LONDON

First published 2019
by Routledge
711 Third Avenue, New York, NY 10017

and by Routledge
2 Park Square, Milton Park, Abingdon, Oxon OX14 4RN

Routledge is an imprint of the Taylor & Francis Group, an informa business

© 2019 Taylor & Francis

Library of Congress Cataloging in Publication Data
Names: Papacharissi, Zizi, editor.
Title: A networked self and human augmentics, artificial intelligence,
    sentience/edited by Zizi Papacharissi.
Description: New York : Routledge, Taylor & Francis Group, 2018. |
    Series: A networked self | Includes bibliographical references.
Identifiers: LCCN 2017059805| ISBN 9781138705920 (hardback) |
    ISBN 9781138705937 (pbk.) | ISBN 9781315202082 (ebk.)
Subjects: LCSH: Human-machine systems. | Artificial intelligence—
    Psychological aspects. | Assistive computer technology. | Self.
Classification: LCC TA167. N35 2018 | DDC 620.8/2—dc23
LC record available at https://lccn.loc.gov/2017059805

ISBN: 978-1-138-70592-0 (hbk)
ISBN: 978-1-138-70593-7 (pbk)
ISBN: 978-1-315-20208-2 (ebk)

Typeset in Bembo
by Apex CoVantage, LLC

To the machines

CONTENTS

FIGURES

CONTRIBUTORS

Courtney Demone is a writer, activist, designer, and sex worker. She has been featured on Mashable, Buzzfeed, the Guardian, and others to talk about queer, transgender, and feminist issues. She is best known for her #DoIHaveBoobsNow project where she examined her loss of male privilege through gender transition by asking the question, "At what point in breast development is it no longer acceptable to be topless?"

Judith Donath is the author of *The Social Machine: Designs for Living Online* (MIT Press, 2014) and has written numerous scholarly and popular articles about identity, privacy, interface design, online communication, and our relationships with not-quite-human others. She created many pioneering online social applications; her work (solo and with her former research group at the MIT Media Lab) has been shown in museums and galleries worldwide. Currently, she is an advisor at Harvard's Berkman-Klein Center and is working on a book about technology, trust, and deception.

Joel Finkelstein is a PhD candidate in psychology and neuroscience at Princeton University and a National Science Foundation graduate fellow. His interests and published work span from optogenetic dissection of brain circuits in disease to statistical dissection of hateful ideologies in online social networks. He currently serves as founder and co-director of the network contagion research institute, which seeks to detect, monitor, and combat the spread of online hate.

Laura Forlano is Associate Professor of Design at the Institute of Design and Affiliated Faculty in the College of Architecture at Illinois Institute of Technology where she is Director of the Critical Futures Lab. Her research is on the intersection between design and technology around themes including ethics, materialities, and futures around topics such as smart cities, distributed work, computational fashion, and medical technologies. Her recent publications include "Posthumanism and Design," "Data Rituals in Intimate Infrastructures," and "Demonstrating and Anticipating in Distributed Design Practices" and she has published in journals including *Design Issues*, *Fibreculture*, and *Digital Culture & Society*. Forlano is co-editor with Marcus Foth, Christine Satchell and Martin Gibbs of *From Social Butterfly to Engaged Citizen* (MIT Press, 2011). She has published chapters for books including editor Mark Shepard's *Sentient City: Ubiquitous Computing, Architecture, and the Future of Urban Space* (MIT Press, 2011). She received her PhD in communications from Columbia University.

Douglas Guilbeault is a PhD student at the Annenberg School for Communication, University of Pennsylvania, where he is a member of the Network Dynamics Group and DiMeNet (Digital Media, Networks, and Political Communication). He is also a member of the Computational Propaganda Project based at the Oxford Internet Institute. His academic research has appeared in the *International Journal of Communication* and the *Journal of Cognitive Linguistics*, and his journalism has appeared in outlets such as *The Atlantic* and *Wired*.

David J. Gunkel is an award-winning educator and author, specializing in the philosophy of technology. He is the author of over seventy scholarly articles published in journals of communication, philosophy, interdisciplinary humanities, and computer science. And he has published nine books, including *Thinking Otherwise: Philosophy, Communication, Technology* (Purdue University Press, 2007), *The Machine Question: Critical Perspectives on AI, Robots, and Ethics* (MIT Press, 2012), *Heidegger and Media* (Polity, 2014), *Of Remixology: Ethics and Aesthetics After Remix* (MIT Press, 2016), and *Robot Rights* (MIT Press, 2018). He has lectured and delivered award-winning papers throughout North and South America and Europe and is the founding co-editor of the *International Journal of Žižek Studies* and the Indiana University Press book series *Digital Game Studies*. He currently holds the position of Distinguished Teaching Professor in the Department of Communication at Northern Illinois University (USA).

Andrea L. Guzman is Assistant Professor of Communication at Northern Illinois University where her research focuses on Human-Machine Communication, people's perceptions of artificial intelligence technologies that function as communicators, and the integration of AI into journalism. She has been integral in spearheading the formation of the HMC community within the communication discipline and is editor of the forthcoming volume *Human-Machine Communication: Rethinking Communication, Technology, & Ourselves*. Guzman's research has been published in *Journalism & Mass Communication Quarterly*, *First Monday*, and *Communication +1*, and has been presented at leading disciplinary and interdisciplinary conferences where it garnered awards at the National Communication Association and the Association for Education in Journalism and Mass Communication. She is a Kopenhaver Center Fellow.

Jessa Lingel is Assistant Professor at the Annenberg School of Communication at the University of Pennsylvania. She received her PhD in communication and information from Rutgers University. She has an MLIS from Pratt Institute and an MA from New York University. She is author of *Digital Countercultures and the Struggle for Community* (MIT Press, 2017) and is currently working on a second book that examines peer-to-peer online marketplaces.

Steve Jones is UIC Distinguished Professor of Communication, Research Associate in the UIC Electronic Visualization Laboratory, Adjunct Professor of Computer Science, and Adjunct Professor of Electronic Media in the School of Art & Design at the University of Illinois at Chicago, and Adjunct Research Professor in the Institute of Communications Research at the University of Illinois at Urbana-Champaign.

Peter Nagy is a postdoctoral research fellow in the Center for Science and the Imagination at Arizona State University. His research interests include the impact of technologies on human identity as well as the public understanding of science. His writing has appeared in the peer-reviewed journals *Social Media + Society*, *Science and Engineering Ethics*, *Journal of Bioethical Inquiry*, and *International Journal of Communication*.

Gina Neff is Senior Research Fellow and Associate Professor at the Oxford Internet Institute and the Department of Sociology at the University of Oxford. She studies innovation, the digital transformation of industries, and how new technologies impact workers and workplaces. Her book *Venture Labor: Work and the Burden of Risk in Innovative Industries* (MIT Press, 2012) about the rise of internet industries in New York City won the 2013 American Sociological Association Communication and Information Technologies Best Book Award. She co-authored *Self-Tracking* (MIT Press, 2016) about the practices and politics of using consumer technologies to track health and other everyday personal metrics. Her ongoing project on large-scale building architecture and construction examines how new information and communication technologies require new ways of working and the challenges of implementing these changes at an industrial scale. She holds a PhD in sociology from Columbia University.

Eleanor Sandry is a lecturer and researcher in Internet Studies at Curtin University and previously a fellow of the Curtin Centre for Culture and Technology. Her research is focused on developing an ethical and pragmatic recognition of, and respect for, otherness and difference in communication, drawing on examples from science and technology, science fiction, and creative arts. She is particularly interested in exploring the communicative and collaborative possibilities of human interactions with non-humanoid robots. She is author of *Robots and Communication* (Palgrave Macmillan, 2015).

Natasha Dow Schüll is Cultural Anthropologist and Associate Professor in the department of Media, Culture, and Communication at New York University. She is the author of *Addiction by Design: Machine Gambling in Las Vegas* (Princeton University Press, 2012), an ethnographic exploration of the relationship between technology design and the experience of addiction. Her current book project, *Keeping Track* (Farrar, Straus, and Giroux, forthcoming), concerns the rise of digital self-tracking technologies and the new modes of introspection and self-governance they engender. She has published numerous articles on the theme of digital media and subjectivity, and her research has been featured in such national media venues as *60 Minutes, The New York Times, The Economist, The Financial Times,* and *The Atlantic.*

Raz Schwartz is a research manager on the Oculus User Experience Research team. Before joining Oculus, he was lead UX researcher on all the location-based products at Facebook. Prior to Facebook, he was a postdoctoral researcher at Cornell Tech and at Rutgers University as well as a research fellow at the Brown Institute for Media Innovation at Columbia Journalism School. His research examines social interactions in an immersive virtual reality environment by applying qualitative and quantitative methods. His work was published in various academic settings and was featured in media outlets such as the *Wall Street Journal, Wired, Rhizome,* and *The Atlantic.*

Tamara Shepherd is Assistant Professor in the Department of Communication, Media, and Film at the University of Calgary. Previously, she held fellowships at the London School of Economics and Political Science and the Ted Rogers School of Information Technology Management at Ryerson University. She studies the feminist political economy of digital culture, looking at policy and labor in social media, mobile technologies, and digital games. She is an editorial board member of *Social Media + Society,* on the executive of the Canadian Communication Association, and her work has been published in *Convergence, First Monday, International Journal of Cultural Studies,* and the *Canadian Journal of Communication.*

William Steptoe is a software engineer and researcher at Oculus and an honorary researcher at University College London. His interests are in virtual and mixed reality. He holds a PhD in Virtual Environments from University College London and has published widely in the field.

Katie Warfield is faculty in the Department of Journalism and Communication at Kwantlen Polytechnic University, and Director of the Visual Media Workshop. Her recent writings have appeared in *Social Media + Society*, *Feminist Media Studies*, *Language and Literacy*, and *Feminist Issues* (6th ed., Pearson Education, 2016). She teaches classes in communication theory, popular culture, discourse theory, media and diversity, and social media. Her research is located at the intersection of post-phenomenology, new materialism, digital literacy, and gender theory.

ACKNOWLEDGMENTS

Let me use this space to acknowledge the support of Jamie Foster, my whiz of a research assistant and presently working toward her PhD on something that will undoubtedly be thoughtful, fun, and provocative. Erica Wetter, here's to your guts and foresight, and not thinking I'm crazy when you asked for a single sequel to the first *A Networked Self*, and I said how about four sequel volumes instead? Mia Moran, I know you will bring many more inspired projects to fruition with your insight.

The rest of you rascals, you know who you are. You are in my heart. I have thanked you personally, many times. I will probably continue to do so. Cheers.

1

INTRODUCTION

Zizi Papacharissi

Every new technology invites its own sets of hopes and fears, raises as many questions as it answers, and resides in its own (false) binary between utopia and dystopia. Inevitably, the dreams and nightmares rendered by the limits of our human imagination revolve around the same theme: Will technology fundamentally alter the essence of what it means to be human? And the answer, despite the countless narratives of anticipation and apprehension is, I find, the same: only if we permit it to do so.

With every technology come new ways of doing things. I am encouraged when we use technology to reimagine and improve our everyday routines, and disappointed when we permit new technology to just reproduce new ways of doing the same old things, across different platforms. So much rests on how we create technologies, the uses that we imagine for them, and the social imaginaries that define the form that these technologies take.

This volume focuses on human augmentics, artificial intelligence, and sentience. It draws inspiration from the work of many luminaries who approach augmented, alternative forms of intelligence and consciousness, and from an interdisciplinary point of view, including (but certainly not limited to) Donna Haraway, Judith Donath, Steve Jones, Mimi Ito, and Judy Wajcman. I invited scholars to contribute their thoughts on how human augmentic technologies and artificial or sentient forms of intelligence can be used to enable, reimagine, and reorganize how we understand our selves, how we conceive the meaning of human, and how we define meaning in our lives.

The starting point of the conversation is this: technology is best put to use when it does not replace but it helps reimagine; when it enhances and enables what is not possible, and complements that which is. The ending point of the conversation lies in the meta, and will be informed by the imagination of the contributors.

This is the concluding volume in the series, and in the spirit of the meta, artistic, and scholarly, license will be given to authors to present their thoughts in the form they see fit. I expect this volume to be unusual, I hope that it will be filled with provocations, and I have designed it to help us reimagine the connection between technology and the (networked, augmented, and enhanced) self.

In "The Robot Dog Fetches for Whom?" Donath considers how humans manage social living among bots and machines crafted to mimic the cues by which we infer each other's emotions, beliefs, and intentions. This modality of imitation is both inviting and disconcerting for humans interacting with bots and machines as fellow sentient beings. As Donath explains, "Robots that resemble familiar beings provide us with a ready-made script for interaction." She ponders whether the extent to which robots function as persuaders, using inherently deceptive imitation to convince us of their sentience, mirrors perhaps our tendency to judge thoughts and experience on performance. Still, robot persuasion "can be used for good. They can be the companion who reminds" us of who we want to be, instead of bystanders who reproduce and acquiesce. But, companionship necessitates more than sentience. It thrives on identity, or, in Donath's words, not just being in someone's presence so as to not be alone, but knowing "who it is that you are with."

Fittingly, Schüll further explores how digital sensors, algorithms, and technologies of self-tracking can evolve out of the restrictive realm of neo-capitalist micromanagement of the body toward an enhanced understanding of *worldly sentience*. In an inspired chapter, that acknowledges the perils of self-tracking, she turns the tables on dominant paradigms and invites us to think along with her, as she explores how:

> rather than dismiss trackers of self-data—as victims of data capitalism and its surveillance apparatus; as life-avoiding and robotically inclined; as digital iterations of the narcissism and depoliticization that Lasch (1991) observed in 1979; or as symptomatic figures of neoliberal subjectivity and its self-mastering, entrepreneurial ethos—we might instead regard them as experimenting with novel forms of subjectivity and sensibility in a digitally networked world.

It is this form of intelligent and informed symbiosis that many of the contributors to this volume advocate, including Forlano, who writes about posthuman capabilities for the networked self. While technologies and discourses around them invite questions of what it means to be human, they rarely afford answers that do not reinforce logics of discrimination and separation; ones that distance human from machine, masculine from feminine, human from animal, thought from action, cognition from emotion and an ever-evolving succession of boundaries that Haraway (1991) sought to advance us away from in her seminal Cyborg Manifesto. In this vein, rather than focusing on the connectivity that the term "networked self" implies, Forlano, using themes of disconnection, disruption, and malfunction, draws attention to false boundaries and provides a language that helps imagine meaning without them.

Speaking of human–machine boundaries, Gunkel, in "Other Things," reminds us that we are in the midst of a robot apocalypse, one perhaps that we have not recognized yet, for it looks far too different from the robot invasion that science fiction narratives suggest. The face of the machine looks far different from that we have imagined, and perhaps too familiar to identify as different from our own, as technologies are produced by humans, and thus, are not as different from our own selves as we might like to think. The challenge of course is to design something that reimagines what we are without blindly reproducing it but also without deftly alienating its creators. Increasingly interactive and autonomous robotic agents invite us, as Gunkel suggests, to think otherwise and beyond relational convention so as to locate *the machine question* within a spectrum "where anything goes and 'everything is permitted.'"

In taking social machines beyond the humanlike other, Sandry makes the case for daring to design machines that do not actualize in our reflection. We imagine and then design machines to mimic and replace us. Perhaps there is a use for these machines. What if, however, a networked self and sense of being fosters communication that thinks with us and beyond us? What if we dare to design machines that do not obey and serve, but disagree with us and challenge our conventions and thinking? Sandry takes us down the path of what things would look like if we thought of chatbots and Twitter bots as social alterities. We often design machines to afford certainty in a world that is unpredictable. We can continue to do so, and we can further create machines that help us reflect and come to terms with that unpredictability.

The literature on artificial intelligence, Guzman aptly points out, as well as technology generally, often speaks in terms of reflections: What aspects of humanity does AI reflect (e.g. Minsky, 1986)? What aspects of humanity does AI distort (e.g. Ekbia, 2008)? How do we view ourselves in the reflection of the machine (e.g. Turkle, 1984)? Refreshingly, Guzman challenges, flips, and ultimately rejects this tendency. She makes the argument that artificial intelligence be viewed as ordinary, rather than an extraordinary and exotically informed capability. Her line of thinking echoes Suchman's (2007) call for acknowledging the reenchantments and mystifications brought on by technological potentialities, so as to be able to reconfigure how we conceive and design technology. In the same vein, Guzman invites us to move beyond reflections of our own being that we see in the machine, so as to perceive what is being reflected back to us. "For 70 years," Guzman argues, "we have conceptualized AI as a technology that inherently challenges our very nature and our humanity" and yet, "when we adopt an approach that situates the study of AI and the self within the context of our lived experiences, a different picture of AI appears," that of AI as ordinary. Ordinary as the technology that drives the mundane, routine, yet exquisite experiences of our everyday lives.

What is the meaning of agency in the posthuman era, ask Neff and Nagy. A concept difficult enough to pin down in the human era, agency is open to a variety of interpretations depending on sociocultural and geopolitical context. The authors suggest the posthuman era invites us to reimagine how we understand agency, and propose we redefine it in terms of how and when people interact with complex technological systems. Drawing from Bandura's social cognitive theory and the concept of symbiosis from biology, they propose an understanding of agency wont to reconcile the human with the machine, thus evolving beyond existing boundaries. The construct of symbiotic agency that they propose permits scholars to evolve out of the debates forming around imposed binaries. Moreover, as the networked self takes form through interaction with human and non-human agents and across spaces actual and imagined, agency, as conventionally understood, may be restricted (e.g. Scholz, 2016). Symbiotic agency, however, may open pathways into advanced forms of an autonomy that may reinvent and re-energize the storytelling project of the self.

Understanding the ways in which autonomy is afforded in technologically enhanced, augmented, or virtual environments is key to designing for symbiotic agency and to enabling the coupling, rather than separation of human and machine (see also Jones, closing chapter to this volume). Schwartz and Steptoe discuss online presentations of the self in virtual and augmented environments to illustrate how avatars can enhance, reinforce, or limit our sense of presence. Through coining the term "immersive VR self," they seek to explain how self, space, and sense of place interact to give shape to embodiment and relational experiences in VR. Focusing on sociological forces that drive embodiment and sense of self in VR is the primary way toward designing environments that integrate human machine agency, intelligence, and autonomy, instead of placing them at odds with each other.

In "Writing the Body of the Paper," Warfield and Demone argue that methods of examining presentations and representations of the self often either read forms or representation as text or examine them as performative curated versions of the self. Seeking to evolve this research paradigm and lessen the distance between representation maker and the researcher who studies them, they work together to introduce three new methods of empirically analyzing how we mediate bodies. First, they suggest *meating* in the story as a way for the researcher to meet the maker in the story told, through visual or other means, in a way that reconciles the distance between the two and provides embodiment of both perspectives in the analysis. Second, they proposed networked intra-viewing as a materialist method of placing emphasis on affect and proximity between participant and researcher, thus advancing interviewing and blending it with the reading and performative practices of the participant and the researcher. Finally, reading horizontally involves the researcher and participant moving together "beyond the humans in the photo to add *dimensionality and depth* to the universe that is operating in the production of the actions in the image." These renegotiated research ethics become increasingly relevant and complex, as we advance to a stage of conducting research with human and non-human entities.

Lingel, in "Clones and Cyborgs," re-examines metaphors of artificial intelligence as ways of alleviating some of the anxiety around AI stemming from the need to delineate non-human modalities of intelligence as artificial. She draws on science-fiction examples and science and technology studies to locate two metaphors that resurface in posthuman renderings of sensemaking: cyborgs and twins. She finds that even though cyborgs tend to dominate how we imagine the future, given their embodied convergence of human and machinic elements, they ultimately fail to capture the potential of a posthuman world. While meaningful in advancing our thinking to a certain stage, they fail to capture the complexities of sociotechnical values. Instead, Lingel suggests that metaphors of twins, clones, and doppelgängers offer advantages in comprehending ownership, subjectivity and agency. Ultimately, metaphors help understand that which is not yet familiar. Perhaps once we advance beyond the point of familiarity, Lingel suggests we will begin to introduce transparency in data doubling, cloning and reproductive practices, and that, eventually, we will consider developing AI with objects that "are less beholden to human mimicry."

In "Human–Bot Ecologies," Guilbeault and Finkelstein explicate the complex symbiosis between biology and selfhood, and by extension, biology and biography through interaction between personal AI bots called *replicas*, developed by the software company Luka. Motivated by an analysis of how the self is an evolving information system augmented via human–bot ecologies and founded upon a rich synthesis of the philosophy, sociology, and psychology of being, they examine how bots are designed to mediate and control the growth of their own selfhood. Central to their analysis are the concepts of *autopoiesis*, that is, self-regenerative systems, and *allopoiesis*, that is, logics that produce products distinct from the self. They propose designing bots as social glia, so as to advance beyond the anthropomorphic culture that has long dominated how we perceive and design for the future of posthuman intelligence.

Shepherd, in "AI, the Persona, and Rights," explores the relevance of the persona rights concept in the context of human augmentics technology, artificial intelligence, and sentience. Drawing from her own work on persona rights as a construct intended to capture the mix of privacy and intellectual property rights that individuals might be entitled to in digital culture, she considers how persona rights might apply to AI, with a view toward regulatory considerations. The construct was created in response to the increasing commodification of user data on social media platforms. Advanced and sentient technology only deepens the salience of understanding persona rights in ways that safeguard individual autonomy while also ascertaining flexibility in how we interact with a variety of human and non-human beings. In this posthuman assemblage, where machines become proxies for the brain, Shepherd asks, where might persona rights be located? She suggests that, in the posthuman era, "letting go of the notion of control over the self as property" might permit us to reconcile the perspectives of free cultural advocates, feminist political philosophers, and indigenous studies into a conceptualization of collective rights that help maintain human dignity and counter colonization of general intelligence by tech companies. As forms of human and machine rendered intelligence come together, it becomes essential to remember that machines are products of and thus reflective of human modalities of intelligence, and to then advance understandings of posthuman persona rights.

In the end, Jones aptly explains that "[m]achines thus shape us and shape our world, while opportunities for us to shape them are few and far between." This does not imply that humans do not possess the ability to shape machines, but that when they do, they often design them in ways that reinforce human biases, thus limiting the cognitive abilities of machine-supported manifestations of intelligence. Jones clarifies that the way out of this is paved by permitting "the processes that technologies engage" to reside in "the realm of thought and not only action," and realizing that "[t]he augmentation is not simply or merely a physical one, whether it is a prosthesis, wearable, augmented reality headset, or some other technological component; it is an augmentation of the mind." Therein lies the answer to the threatening potential technology pioneers and everyday users associate with human augmentics and AI. Technological advancement does not de facto run counter to human progress. Common sense as it may sound, it is the principles that guide technological advancement that frequently contradict the essence of human nature. And, ironically, those principles are put in place and reinforced by humans.

We are human; we must also remember to be humane when we design our future. Technology is generated by humans, so any threat embedded in its design is a result of our human nature. Technology is human and humans are technology. Any misstep toward our future will not be technologically rendered; it will be human-made. Any flaw in the design of technology is connected to the limitations of human ways of thinking.

In a playful encounter with performance artist's Stelarc's (2017) *Prosthetic Head*, a three-dimensional simulacrum of the artist's own, capable of responding to queries typed on a keyboard, Lucy Suchman carried on the following conversation:

LS: (*types*) "Do you have feelings?"
HEAD: As a robot I do not have any human emotion. My apparent emotional responses are simply algorithms designed to make me easier to interact with.
LS: (*types*) "Do you think robots should have emotions?"
HEAD: That depends on what you mean by thinking.

(*Suchman, 2007*)

I pursued curation of this volume because I was interested in exploring what it means for the networked self to think, feel, and act. The contributors to this volume have generously delved into deep conversation about the meaning of these three words and have also taken on the task, here and in their other work, of reimagining how we define thinking, feeling, and acting.

By focusing on the theme of augmentics, I aimed at directing our attention to ways in which technologies expand our capabilities. Having read what contributing authors have to say on the matter, I find, once again, that we are restricted by language. I am no longer sure that augmentics is the ideal vocabulary with which to describe the capabilities of technologies to come. I do not contend that algorithmic logic does not amplify our ability to access and process information. Of course, prosthetics expand our competence. Naturally wearables may improve our faculties. That is not all they do, however, and if we imagine and then design technology as something that augments or enhances our human powers, then we are bound to wind up with bizarre superhero complexes, to say the least, and grossly underestimate *techne*, more importantly. It is important to understand that augmentics, for lack of a better word, will make us different. We will not become better, superior, or more advanced—that will be a function of how we put technology to use, and ultimately, a call that will be forever subjective. We will, or rather, we have the opportunity to, become different, and with this opportunity comes another possibility, perhaps an obligation: to reimagine how we do things; to do things differently; to not fall into the trap of reproducing what we already have.

Sentience and artificial intelligence presented two further directions I wanted our contributors to think out loud about, and I am grateful for the unusual paths they have taken in articulating these thoughts. I have always been puzzled by the term "artificial intelligence" and the ease with which it is employed to describe the capabilities of non-human agents. Simply put, what is so artificial about it? If our goal is to design a form of intelligence that is autonomous, then labeling it artificial seems modest, at best, and judging it by comparing it to our own strikes me as rather presumptuous. Perhaps it is time to think toward the formation a new reason (Stiegler, 2016). Sentience is no different. We have long been bound by the belief that non-human agents do not feel. As many have shown, with Haraway leading the way, there are many modes of feeling, different from our own. Authors contributing to this volume embrace forms of intelligence and sentience that evolve beyond boundaries imposed by humans. They lead the way to accepting intelligence and sentience as open-ended ways of describing a wide range of aptitude.

In closing, I will say, once again, that we are limited only by our own ways of organizing our world. At the very present, we are limited by language. Artificial intelligence, augmentics, and sentience are prevalent terms, but may not be the best ones to describe the world we are creating. Intelligence may even be inadequate, for it carries its own sociocultural biases, metrics, and assumptions. We are all intelligent in different ways and we are also not consistently intelligent. The human brain is a machine, with its own code, chemicals, and viruses. Similarly, sentience comes with its own prescribed vocabulary of emotions that science has long discarded. We describe ourselves as happy, sad, or energized, when these are all labels that we have been conditioned to ascribe to certain affects. We recognize these as deeper emotions despite these typically being fleeting states of being. In designing our future, we might reorganize our language; our code for semantically interpreting our states of being.

This anthology is driven by an emphasis on storytelling, so it is fitting that I return to the words we use to tell our stories. Signs, be they spoken, written, symbols or data, reflect the storytelling conventions of each era. These conventions suggest an orality for each epoch; a dominant way of communicating. Ong (1982) artfully explained how visual depictions of knowledge, for instance, came to be considered superior to previous written or verbal forms of preserving and sharing knowledge. He famously distinguished between a *primary orality*—where the form of sharing stories organically evolves as knowledge is sounded out through the spoken word to listening publics who are aware that this evolving continuity is part of the essence of communicating knowledge—and a *secondary one*—where the deliberate spontaneity of a writing and print culture mutes out the subjectivity inherent in voicing these stories, in favor of generating "technologized," "permanent," and thus "silent" stories that distance. Elsewhere, I have argued that an example of this form of distance is encountered within the paradigm of journalistic objectivity, which requires narrators to establish objective distance from the story so as to ascertain accuracy and thus electronically reproduce and share verified information (Papacharissi, 2015).

Drawing from Ong's (1982) work, I have understood the present era as one undergoing a transition toward a digital orality: a curious blend of sounded out and muted, spoken and written, ephemeral and permanent, editable and remixable, spontaneous yet always datafied storytelling practices (see Papacharissi, 2015; Gunkel, 2016). Technologies of the present afford us, among other things, a way to reconcile the distance between a primary and secondary orality, and a way to reinvent how we tell stories, how those stories turn into knowledge, how that knowledge is preserved and shared, and ultimately, what constitutes knowledge. As the storytelling canon of our era evolves, a new language will take shape for the networked self, and I hope it is as different from the languages that we invented in the past as it can be, while also bringing us closer to the technology we create. Perhaps, to paraphrase de Certeau (1984), this new language will no longer privilege speaking. The language of the networked self is code; however code may be defined in the future.

References

de Certeau, M. (1984). *The practice of everyday life*. Berkeley, CA: University of California Press.

Ekbia, H. R. (2008). *Artificial dreams: The quest for non-biological intelligence*. New York: Cambridge University Press.

Gunkel, D. (2016). *Of remixology: Ethics and aesthetics after remix*. Cambridge, MA: MIT Press.

Haraway, D. J. (1991). A cyborg manifesto: Science, technology and socialist-feminism in the late twentieth century. In *Simians, cyborgs and women: The reinvention of nature* (pp. 149–181). New York: Routledge.

Minsky, H. P. (1986). *Stabilizing an unstable economy.* New Haven, CT: Yale University Press.

Ong, W. J. (1982). *Orality and literacy: The technologizing of the world.* New York: Routledge.

Papacharissi, Z. (2015). The unbearable lightness of information and the impossible gravitas of knowledge: Big Data and the makings of a digital orality. *Media, Culture & Society,* 1–6. doi: 10.1177/0163443715594103

Scholz, T. (2016). *Uber-Worked and Underpaid. How Workers are Disrupting the Digital Economy.* Cambridge, UK: Polity Press.

Stelarc (2017) *Prosthetic head.* http://stelarc.org/?catID=20241

Stiegler, B. (2016). The formation of new reason: Seven proposals for the renewal of education. In D. Barney, G. Coleman, C. Ross, J. Stere, & T. Tembeck (Eds.), *The participatory condition in the digital age* (pp. 269–284). Minneapolis, MI: University of Minnesota Press.

Suchman, L. (2007). *Human-machine reconfigurations: Plans and situated actions.* Cambridge: Cambridge University Press.

Turkle, S. (1984). *The second self.* New York: Simon & Schuster.

2

THE ROBOT DOG FETCHES FOR WHOM?

Judith Donath

A Beautiful World

In the picture, a boy sits with his dog at the edge of a beautiful and serene lake. Under the blue sky, the far shore is visible, with a short wide beach and green woods behind it. The boy wears jeans and an orange sweatshirt; the dog has a proper collar. The boy is human. The dog is a shiny, metallic robot.

The picture is an ad for OppenheimerFunds, and the copy is headlined "Invest in a Beautiful World." The reactions it elicits are mixed. Some people viewed the scene as "heartwarming," a vision of a pristine future where technology is our best friend. Others thought the dog was disturbing, a "weapon-like creature" that provoked a "visceral sense of unreality and horror."

Like many ads, it is deliberately a bit confusing and provocative—crafted to make you pay attention, to think about it, maybe to get involved in a discussion online about what it really means—all to etch the brand name and the general aura of the campaign into your mind. This is what ad companies do—they are masters at influence and persuasion.

I can see why some people might find the scene heartwarming—it is a lovely setting. And though you only see the boy's back, he evokes a very all-American sort of innocent, free youth, with his denim jeans, his cotton sweatshirt—and his loyal companion dog. Sitting down, they are roughly the same size. The dog's head is held high, his ears a bit cocked forward—he is alert, protective. A boy and his dog: innocent, independent.

The sponsoring company claims to have intended this positive interpretation. We have "anchored our message on optimism," the chief marketing officer says of the campaign. "Once you look to the long term and expand your view, as we do at OppenheimerFunds, the world reveals itself to new opportunities that others may have missed." That is not how it feels to me. The serenity of the picture contrasts vividly with the scenes in which we typically encounter robot companions, such as post-apocalyptic movies in which WWIII has scorched the earth bare. There is a fear in the back of our minds that if we look far enough into the future to see fully functioning artificial pals, we will also be looking at a world fully denuded of a natural beauty: at worst a nuclear wasteland; at best a desiccated, over-built, paved and strip-malled (strip-mauled) earth. Is the scene here as pristine and natural as it seems, or is it, like the robot dog, a simulation of the organic? The dog appears to be the boy's loyal companion—but is he? Who programmed the dog? Whose orders does he follow? Does the boy lead him or does he shepherd the boy? Are their travels the secret adventures of a boy and his dog—or does the robot report back about where they go, who they see, what the boy confides in him? If so, to whom does he report?

The Shiny New Best Friend

The image of the boy and his robot dog is futuristic but also plausible. While robot companions are still in their clunky infancy, we already see evidence that man's best friend can be—or be replaced by—a robot dog.

Robot pet dogs are numerous enough that we now have "10 best Robot Dogs in 2017" lists. One of the earliest and most well loved was Sony's AIBO. Produced in the early 2000s, AIBO was an expensive "toy"—well over $1000—but quite popular: in 1999, all 3000 units of a production run allocated for Japan were bought in 17 seconds (Johnstone, 2000).

The roots of AIBO's appeal lay in its genesis as a research project. Unlike many cheaper objects advertised as robot dogs, but better described as animatronic toys, AIBO was not the product of a marketing division, but was created by highly regarded scientists, engineers, and artists interested in advancing knowledge of robot locomotion, autonomy, and human interaction who set themselves the challenge of making a robot that would engage people over a significant period of time.

Sony Corporation wanted to explore the potential of robotics and artificial intelligence for the home. AI expert Masahiro Fujita suggested developing a robot framed as a "pet," in part because it was more feasible than other, more utilitarian applications. It would need neither sophisticated natural language skill nor the ability to carry out challenging physical tasks; it would not be relied up on for critical tasks (Fujita, 2004). Instead, the robot—which came to be known as AIBO (Artificial Intelligence Bot)—would be clever.

Fujita, the lead inventor, argued that in order to be interesting the robot would need to have complex movements and behaviors; thus AIBO was built with a powerful controller, a variety of sensors and activators, and a sophisticated behavioral program (Fujita, 2004; Fujita & Kitano, 1998). AIBOs' behaviors, based on an ethological model of motivation, were neither so predictable as to be easy to figure out and thus machinelike, but not so unpredictable as to seem random and unmotivated. They were programmed to have internal "emotional states," which interacted with external stimuli to trigger various actions. Owners could reinforce different behaviors, and as an AIBO learned over time, it developed an increasingly individual personality. If you approached an AIBO and reached your hand toward it, it was likely to give you its paw to shake—but it also might hesitate, ignore you, or do something entirely different, depending on recent activity that affected its "mood state," its past experiences, and so on (Arkin et al., 2003).

AIBO was also cute, with a rounded puppy-like body, ears that flapped and a tail that could wag. Yet it was also distinctly mechanical looking, a design choice aimed to avoid the "uncanny valley" (Mori, 2012) and to lower people's expectations, thus by contrast increasing the impact of its relatively life-like movements and autonomous behaviors.

Most importantly, it was designed to appear sentient. Even the timing of its responses was crafted to enhance this illusion: it would respond rapidly to something such as a loud sound, but would pause, as if deliberating, before responding to certain other stimuli.

The depth of many AIBO owners' attachment to their robots attests to the success of this project. Many felt great affection for their little robots, describing them as beloved pets, even as family members. Some AIBO owners reported playing with their robot dog for about an hour on a typical day.

Yet, as in any corporate endeavor, the researchers were not the ultimate determiners of AIBO's fate. Although the robot dogs sold well, had deeply loyal owners, and generated enthusiastic publicity, in 2006, Sony, under new management, announced the end of the AIBO product line. Research and development stopped immediately, and the remaining inventory was sold, though the company did provide support and spare parts for several years. When that stopped in 2014, private repair businesses opened to provide "veterinary" care for the aging robots. One such robot veterinarian, Hiroshi Funabashi, a former Sony engineer, said "The word 'repair' doesn't fit here. For those who keep AIBOs, they are nothing like home appliances. It's obvious they think their (robotic pet) is a family member" (AFP-JIJI, 2015).

Eventually, however, necessary spare parts became scarce. Replacements were available only from "deceased" robots "who become donors for organ transplantation, but only once the proper respects have been paid" (AFP-JIJI, 2015).

In 2015 a funeral was held at the 450-year-old Kofuku-ji Temple in Isumi, Chiba Prefecture, Japan. Nineteen no-longer-functioning AIBO robot dogs lay on the altar, each identified with a tag showing their hometown and the family to whom they belonged. The head priest, Bungen Oi, said a prayer for them; the ritual, he said, would allow their souls to pass from their bodies (AFP-JIJI, 2015).

Reactions to these funerals varied. Some saw them as heartwarming expressions of empathy with AIBO owners, some thought they were silly, and others were offended by them, seeing them as a perverse mockery of rites that should be reserved only for humans, or at least, for once-living beings.

Almost Mutual Affection

AIBO owners knew that their robots were not living beings. But people grow quite attached to things—and have been doing so since long before the advent of robot dogs (Turkle, 2011). We grow attached to things that we think of as individuals, such as the dolls and stuffed toy animals that a child names and imagines personalities for and adventures with.

We grow attached to things with which we work or play: cars, bikes, espresso machines. This is especially true when they require skill to use (think of a chef and her knives) or when they do not work perfectly, when they need to be coaxed and handled just so (Donath, 2004, 2007; Kaplan, 2000). (Arguably, the decline of car culture among teens from the days when getting one's driver's license was a greatly anticipated rite of passage to today when many are indifferent to driving is a result of the car's transformation from a finicky machine open to tinkering, to a high-tech but boringly predictable and closed product.)

We grow attached to things that we have altered and that have conformed to us: worn-in jeans, well-read books, the table marked by a generation of dinners. Through our interactions, these items, once anonymous commodities, have become both individual, distinct from their initial counterparts, and personal, incorporating a bit of ourselves and our history (Donath, 2011; Kopytoff, 1986).

It is not surprising that AIBO owners grew so very attached to their cute robot dogs that ran about and learned new tricks. The AIBOs featured all the elements that induce attachment to objects—and much more. It was designed to appear as if it was sentient,[1] with its manufactured simulations of thoughtful pauses and other such tricks; it was made and marketed to resemble a dog, an animal known for its loyalty and love for its owners; it learned new tricks and habits, changing in response to its owner's actions.

Is this desirable? Do we want to be building—and buying—machines designed to create such emotional attachments?

A strong argument, dating back to the first computational agent that people treated as a sentient being, says no, this affection is potentially dehumanizing. That agent was the chatbot ELIZA, a simple parsing program which in its initial (and only) role followed a script in which it mimicked a psychologist who, along with a few stock phrases, mostly echoed the users' words back to them in question form (Weizenbaum, 1966). Its creator, MIT professor Joseph Weizenbaum, expected that people would see it as proof that humanlike conversational ability was not a reliable signal of underlying intelligence; he was profoundly dismayed to see instead that people enthusiastically embraced its therapeutic potential (Weizenbaum, 1977). For Weizenbaum, this easy dismissal of the significance of interacting with a real person, with real feelings and reactions, demonstrated a dangerous indifference to the importance of empathic human ties (Weizenbaum, 1976).

Fundamentally, society functions because we care about how other people think and feel. Our desire that others think well of us motivates us to act pro-socially, as does our ability to empathize with others (at least those we are close to, including our pets): their happiness becomes our happiness, their pain, our pain (Eisenberg & Miller, 1987). When we care about another, we want to make that person or animal happy. But what if that other does not actually feel? Think about playing fetch with a robot dog. People, for the most part, do not play fetch with other people: it simply is not inherently much fun. The entire point of playing fetch with a dog is that it is something you do for the dog's sake; it is the dog that really likes playing fetch—you like it because he likes it (Horow-itz & Hecht, 2014; Marinelli et al., 2007). If he does not like it because he is a robot—and, while he acts as if he is enjoying it, in fact he does not actually enjoy playing fetch (or anything else) because he is not sentient, he is a machine and does not experience emotions—then why play fetch with a robot dog? Psy-chologist and historian of science Sherry Turkle has conducted several studies of people and their relationships with autonomous beings (Turkle, 2007). One of her deepest concerns is the enticing ease of these pseudo-relationships. A robot dog need never pee on the carpet; a robot boyfriend need never stay out late, flirt with your friends or forget your birthday. And indeed, many AIBO owners cited convenience as a significant reason why they chose a robot over a real dog: proclaiming to love their pet, they also liked being able to turn it off when they went on vacation (Friedman et al., 2003). The dystopian interpretation predicts that robot-enabled narcissism will destroy our patience for the messy give-and-take of organic relationships, spoiling us with the coddling and convenience of synthetic companions.

Even if the future is not so bleak, our relationships will be different: "Ulti-mately, the question is not whether children will love their robotic pets more than their animal pets,[2] but rather, what loving will come to mean" (Turkle, 2007).

The simple but addictively compelling Tamagotchi key-chain pet has been the most popular artificial "creature" thus far. It has been quite effective in persuad-ing people to devote considerable and often inconvenient time to taking care of it—pressing the proper button in response to its unpredictable but urgent need to be fed, cleaned, or entertained (Donath, 2004; Kaplan, 2000). Neglect it, and the Tamagotchi dies (or in some gentler versions, runs away). Imagine a child at a family gathering who is ignoring the conversation, distracted by a Tamagotchi. The parent who views it as just a toy might say: "Put that away. I don't care if your Tamagotchi dies; it is not real, while this is your flesh and blood grandmother, these are actual people you need to pay attention to and be present for." In con-trast, the parent who sees it as practice for care-taking might say: "Well, I don't want to teach you to be cold and heartless. The Tamagotchi is teaching you to be nurturing and responsible, and we want to encourage that." One can argue that encouraging nurturance is good, regardless of the capacity of the recipient to feel it. Compassion is not a finite good that you use up when you care for something; instead, it is a practice that grows stronger with use.

People vary in their propensity toward anthropomorphism, that is, in their tendency to perceive humanlike intentions, cognition, and emotion in other animals and things (or, similarly, zoomorphism, the tendency to perceive animal qualities in inanimate objects). The greater your anthropomorphic tendencies, the more social and emotional your perceptions of and reactions to social robots will be (Darling, 2017). And anthropomorphic tendencies do not lead people to devalue real humans and animals; on the contrary, anthropomorphism has been linked to deeper concern for the environment (Tam et al., 2013) and decreased consumption: valuing objects more highly and taking responsibility for them discourages casually discarding (or acquiring) them.

Differences in anthropomorphic tendencies have both biological and cultural roots (Waytz et al., 2014). People with autism generally have little tendency to ascribe humanlike qualities to non-human agents; among the general population, differences among individuals have been found to correspond to neurological differences (Cullen et al., 2014). Cultural differences also influence the degree to which one perceives human traits in animals or personality and intent in inanimate objects. While Judeo-Christian beliefs specifically forbid worshipping "idols," Shinto practice sees spiritual essences residing in plants, animals, rivers, rocks, and tools.

The AIBO funerals, which from a Western perspective seemed strange, even absurd, look different when viewed in the context of related rituals. *Kuyō* is the Shinto-based Japanese practice of performing farewell rites to express gratitude to useful objects. For example, the *Hari-Kuyō* festival commemorates broken needles (and celebrates the women who had sewn with them), and there are similar rituals performed by musicians for instrument strings, Kabuki actors for fans, near-sighted people for glasses, and so on (Kretschmer, 2000). It is an approach that is finding resonance in the West. Marie Kondo's book *The Life-saving Magic of Tidying Up* provides advice for de-cluttering your home and life, with a central rule being that when you discard something, you thank it sincerely for the service it had provided (Kondo, 2014); millions of copies of her books have sold around the world and she has been named in *Time* magazine's annual most influential people list.

Anthropomorphic by Design

This does not mean that the concerns Weizenbaum and Turkle had about people's relationships with social agents and robots are unfounded. There is a subtle but important distinction between traditional anthropomorphized objects and today's artificial interactive beings.

The objects thanked in the *Kuyō* ceremonies are being acknowledged for what they are and the service that they performed in that capacity. While they may be anthropomorphically perceived as having personality and intentions, they were not deliberately crafted to create that impression. Rather, their characterization as animate or sentient beings is bestowed by the objects' users, emerging from their experience and interactions with the needle, the fan, the eyeglasses, and so on.

The robots that disturb Weizenbaum and Turkle are different: they are designed to seem conscious. Indeed, it is very difficult *not* to think of them as thinking, feeling beings (Melson et al., 2009; Pogue, 2001). To perceive agency in traditional objects requires active imagination; perceiving it in chatbots, social robots, and so on needs only passive acquiescence to this designed manipulation.

Neither the needle nor the AIBO is actually sentient, but the latter's inexorable emotional tug adds a new element to the question of what our responsibility to these objects is. While Marie Kondo suggests that you thank things for their service, she says this in the context of advising you to be ruthless in discarding them. The robot dog that wags its tail fetchingly at you as it looks up to you with its big round eyes cannot be disposed of so easily (Archer, 1996). And while it would in fact be no more consciously aware of being discarded than a needle is, do we want to inure people to that instinct for compassion?

The motivations of a companion robot's designers matter. AIBO's inventors were AI and robotics researchers—they wanted to advance development in these fields. Working at Sony, they also needed to sell their ideas to top management: they needed to make something that people would buy. Fujita, AIBO's primary inventor, had suggested an entertainment robot framed as a "pet" as a project that would satisfy the interests of both management and the researchers. It is important to note that they were not attempting to make a robot that would sell things or persuade it owners to do anything other than care for and play with it.

One reason people love animals is that they are innocent dependents (Archer, 1996). Your dog may win some extra treats with big brown pleading eyes, and I find it impossible to resist my cat when he carries a toy to my desk, dropping it with little meows that seem to say "I've brought you this gift, and now we must play with it." But their motives are straightforward. They do not pretend to like someone in order to fleece them or flatter their way up the ladder. They are not salespeople who feign friendliness and admiration to sell you a dress they secretly think is hideous and overpriced.

AIBO was a similarly "innocent" robot, created by researchers who wanted to understand how to make something entertaining and likeable. It was manipulative in the sense that it was designed to give the illusion of a sentience that it did not have—but it did not have ulterior motives: it was not designed to pry out secrets, or to persuade its owners to do anything beyond liking it and playing with it.

Other robots are not necessarily so innocent.

Persuasive Robots

In the next few years we are likely to face a generation of social robots designed to persuade us. Some will be motivating us at our own request—weight-loss and exercise cheerleaders, for example. But most will be directed by others, whose intentions for influencing us are guided not by an altruistic concern with our well-being, but by goals such as getting us to buy more of what they are selling. Such goals are often not in our own best interest.

There are three questions we want to address to consider this more deeply:

- How persuasive will robots be?
- How likely is it that they are going to try to sell us things and how extensively?
- Is this a problem and if so, why?

The desire to please others, to create a desired impression in their eyes, is the basis of persuasion. We care deeply about what others—including anthropomorphized objects—think of us. Once we think of something as having agency—having a mind—we ascribe to it the ability to form impressions of us and to judge us. This gives it the power to influence us, for when we think that we are being observed and judged, we change our behavior to make a more favorable impression.

Thus, the design of social robots to mimic human or animal behaviors—to elicit anthropomorphic reactions—makes it very likely that these machines will be influential partners to the people who befriend them.

An early study of the effect of humanlike computer interfaces demonstrated quite vividly that people present themselves differently to an easily anthropomorphized interface than to a machinelike one (Sproull et al., 1996). Asked questions about themselves (ostensibly as part of a career guidance survey) they were more honest with a purely text interface; when the interface featured a human voice and face, they instead strove to present themselves more positively.

Social robots could be quite subtle in their persuasion. They are objects that have a long-term relationship with people, so their message can be slowly laid out over time. It need not make any overtly commercial or partisan remarks, but rather establish itself as an entity that you want to please—one with certain tastes. These tastes could be manifest in comments it makes, how it introduces news stories, presents music and movie choices, and so on.

Indeed, there are many strategies for making robots increasingly persuasive. Researchers in this active and growing field investigate techniques such as the effect of changing vocal pitch (Niculescu et al., 2013), gender (Siegel et al., 2009), and varying gaze and gesture (Ham et al., 2011).

Papers published in this field frequently emphasize that the goal of the research is to make social robots that help people achieve personal goals. Cynthia Breazeal, one of the field's leaders, describes the goal of her work as creating a robot that will "sustain its user in a highly personalized, adaptive, long-term relationship through social and emotional engagement—in the spirit of a technological 'Jiminy Cricket' that provides the right message, at the right time, in the right way to gently nudge its user to make smarter decisions" (Breazeal, 2011). Ham and colleagues note: "Whether it is actual behavior change (e.g. help the human walk better, or take her pills), or a more cognitive effect like attitude change (inform the human about danger), or even changes in cognitive processing (help the human learn better), effective [sic] influencing humans is fundamental to developing successful social robots" (Ham et al., 2011). But there is nothing inherent in the persuasive techniques that earmark them for such beneficial uses.

Amazon's Alexa, introduced in 2015, is an artificial personal assistant that communicates by voice. She does not look like a classic robot: her physical form is just a simple round speaker. But the voice is embodied, personal, female, and faintly alluring. So while Alexa is not a robot in the mechanical sense, she is an artificially intelligent-seeming being that inhabits your house. Alexa finds information, plays music, checks the weather, controls smart-home devices—and, of course, helps you buy things from Amazon.

Alexa makes shopping seamless. If you think you need something, just say out loud "Alexa, order some salsa," or an umbrella, a book, a full-size gas grill. Alexa is a magic genie; your wish is her command. Invoke her, and in a day or two, the item you desired appears at your door. Business analysts predict that Alexa could bring Amazon over $11 billion in revenue by 2020—$4 billion in sales of the device itself and $7 billion in commercial transactions made via this agent (Kim, 2016).

And that is just today's Alexa, the voice in the little round speaker. It is still in its infancy, putting through basic orders and re-orders. But its skills, as the apps for it are known, are growing. As I write this chapter, Amazon announced a new version of the speaker, with a built-in camera. You put the device in your bedroom, and Alexa will give you advice about what to wear—and what you need to buy to perfect your outfit.

Devices such as Alexa are always listening, at least for the keyword that puts them in full listening mode. Even without "hearing" everything that goes on in a house, a digital assistant learns what music you like, what information you need, what circumstances prompt you to ask for jokes, advice, cocktail recipes, or fever remedies—a series of snippets that together portray the inhabitants' habits, preferences, and needs. (And I expect that soon trusted assistants will be given permission to listen at all times—people have proved to be quite willing to give applications sweeping permissions in exchange for "a better user experience.")

Your relationship with Alexa is not like that with other possessions, or even pets. If you bought an AIBO, it became your robot dog, your artificial pet. You were its owner, and all its parts, for better or worse, were yours. You are not, however, Alexa's owner: Alexa has customers. Furthermore, Alexa is not acting solo. Alexa's brain is not in the little round Echo speaker; Alexa's head is in the Cloud—where your current request is added to the vast dossier of searches and purchases and past queries, the vividly detailed portrait of you.

For the casual user, much about Alexa is opaque, from her actual abilities to the corporate goals that guide her. Personal robots such as AIBO the beloved pet and Alexa the trusted assistant are designed to encourage people to develop relationships with them as if they were sentient. We want to keep them happy and we want them to think well of us: two desires that enable them to be quite influential. A message from someone we care about—and trust—carries weight that a typical ad does not. Is that trust warranted?

Mimicking Trustworthiness

We can infer much about the inner state and capabilities of living creatures from their outward appearance. Upon seeing a dog, for instance, we expect it to understand (or be able to learn) a few commands—sit, fetch, lie down, and so on. We might expect it can read our emotions and respond based on circumstance and personality: we are taught not to show fear to an aggressive dog lest it attack, while an emotional support dog knows to comfort its nervous owner. We also have expectations about the limits of its abilities: we discuss confidential material in front of dogs, cats, and infants, knowing they can neither understand nor repeat it.

Robots that resemble familiar beings provide us with ready-made scripts for interaction. People easily understood that an AIBO, like the nice dog it invoked, would shake your hand with its paw, wag its tail when happy, and fetch the ball that you tossed.

But robots are more cryptic than living beings: their outward appearance is generally a poor guide to their actual abilities and program. A cute robotic kitten or baby doll can be equipped with the ability to process natural language or to transmit everything it hears to unknown others just as easily as one that looks like an adult human or an industrial machine. A robot that asks about your feelings because it is running a helpful therapeutic program may appear identical to and ask the same questions as one programmed to assess your tastes and vulnerabilities for a consumer marketing system or a government agency. If a robot does behave in the manner its outward appearance suggests, it is because its creator chose for it to do so. It is up to the robot's makers to decide how much they want their creation to internally and behaviorally replicate the creature that it mimics.

Social robots can mimic the behaviors and appearances that lead us to trust another living being. They may resemble something we find trustworthy, such as a faithful dog or childlike doll. They may mimic expressions, such as a calm demeanor and direct gaze, that lead us to trust another person. In their original context, these cues are reasonably reliable signals of trustworthiness: while not infallible, their inherent connection to underlying cognitive and affective processes grounds their credibility. But there is no such grounding when they are mimicked in an artificial being's design. A robot designed to elicit trust is not necessarily one designed to be worthy of that trust.

Will we remember, as we populate our homes with robot companions, that their outward appearance may imply intentions far removed from their actual goals (Calo, 2009)?

Robot Campaigners

One such covert goal is commercial persuasion: robots that establish a trusting relationship with a person, then use that relationship to market purchases and build consumer desires. One of the most popular domestic bots today is Alexa, brought to you by amazon.com, the world's largest online retailer—a provenance that supports the prediction that many household robots will seek profitability for their parent company by becoming a marketing medium, one that can both sell to and gather extensive information about you, the user.

Living things evolve to survive in a particular niche. Domestic animals—such as real dogs—have evolved to survive in a niche defined by human needs and tastes. Commercial products can also be said to "evolve" (though with deliberate design and without the genes) and need to survive in often harsh and profit-seeking corporate environments.

From a design perspective, the AIBOs were successful. People became very attached to them: they played with them for hours and spoke of the robots as pets and family members. But from a corporate perspective, the AIBO line was less thrilling. Although thousands were sold, they were so expensive to produce that this was not enough for them to be profitable. The AIBO was invented and thrived in a period of corporate wealth and generous research funding. But when Sony faced financial trouble, the new CEO, seeking to eliminate any unnecessary spending, ended the project. The individual AIBOs were left to each succumb to broken and eventually unfixable motors and gears.

Innumerable products have met similar fates. People want them, but do not want to pay the price needed to make them profitable enough. In 1836 the French newspaper *La Presse* found that by running paid advertisements it could significantly lower its price, grow its readership, and become more profitable. Since then we have seen that adding advertising—selling the attention of your readers, users, or customers to others who want to get a message to them—is a potently effective way to make something—a magazine, a bus, an Instagram feed—profitable (Wu, 2016).

In the earliest days of the web, few foresaw the enormous role that advertising would play. Though ads were present almost from the beginning, they initially seemed like a small side feature in the excitement about the new medium, where people were publishing their photos, their know-how, and their musings on life, and vast troves of knowledge were being woven together.

Today, the tentacles of web advertising are everywhere. Unlike print and television ads, online ads do not just feed you enticing images and information—they track you through a surveillance ecosystem in which the advertisements a) fuel the need to build detailed dossiers about everyone, in order to better serve targeted ads to them (targeted both to what they are likely to want and, in some cases, in a strategy to be the most persuasive to that person) and b) are themselves a data-gathering technology for those dossiers by tracking people as they move about the web. A web ad may superficially resemble its counterpart in print form, but the underlying technology—the network of trackers, the personalization—makes it a vastly more powerful and more insidious form. And while personalization is still rather primitive, the data that has been gathered about many of us is detailed and extensive, paving the way for more sophisticated models to predict wants and vulnerabilities.

In these early days of social robots, few foresee that they will become a powerful and invasive advertising medium. But it is likely that they will, and we should consider that prediction now, if only to prepare for (or head off) this likely future (Hartzog, 2015). Like web ads, they will both surveil and market. Also, like web advertising, robot (or agent)-based advertising will not be just a continuation of the old. The data that a domestic robot can gather is potentially far more intimate and detailed than what a web-based ad-network can find. More radically, it is their ability to market—the persuasive capabilities of personable, anthropomorphic companions—that will put them in an entirely new category. Recent studies, for example, show that robot social feedback—even with a rather primitive robot—is more effectively persuasive than factual feedback (Ham & Midden, 2014).

And future robots will not only be more sophisticated, there will be more of them. A scenario to consider is one where there are multiple cooperating robots. What are the social dynamics when you are among a clique of social bots? Think of the social pressure once you have three Alexas in the room, and they are all chatting and friendly, and they all really like this political candidate, and you—well, you are not sure. But you like *them*, and when you express your doubts, they glance at each other, and you wonder if they had been talking about you among themselves, and then they look at you, a bit disappointed.

We need to better understand the potential effectiveness of such marketing, whether consumer or political, before we willingly populate our homes and workspaces with persuasive robots, whose minds have been shaped and are controlled by interests far from ourselves.

> "Ginny!" said Mr. Weasley, flabbergasted. "Haven't I taught you anything? What have I always told you? Never trust anything that can think for itself if you can't see where it keeps its brain?"
>
> (J. K. Rowling, *Harry Potter and the Chamber of Secrets*)

Thinking for Whom?

I would like to be optimistic about these robots. As an engineer/artist/designer, I think they present an array of fascinating research challenges, from enabling them able to learn and to infer information in informal settings to designing the nuances of their social interactions.

They can be persuasive—and that ability can be used for good. They can be the companion who reminds you of who you want to be.

Our fascination with robots can itself be a cause for optimism. The ability to anthropomorphize, to perceive sentience and spirit in the objects around us, does not mean devaluing living things as much as it means bringing a broader collection of things into the sphere of our empathy and concern. Environmental groups in Japan are pushing for a revival of Shinto customs revering the spirits in non-living things as a way of getting people to consume and discard more thoughtfully and minimally (Rots, 2015).

But to get to that future, we really need to invest in a beautiful world. We need to know what our robot companions are thinking, and for whom are they thinking.

A boy sits at the edge of a lake with his real dog. What is the dog thinking? We do not really know. We can guess: he is watching for birds, hoping for a snack, and thinking about the feel of the summer breeze. Does he—can he—think about the trip back home or last night's dinner or last winter's jogs in the snow? We do not know. But there are some things we do know he is *not* thinking. He is not thinking about telling the boy to buy a new bike or go to church more often.

A boy sits at the edge of a lake with his robot dog. What role does the dog play? Was he, like the AIBO, designed to be as good a companion as possible? Or is this handsome, expensive robot affordable to ordinary families because in fact he is a persuasive conduit for the sponsors who underwrite his cost? He knows a lot about the boy—does he convey this information to the boy's parents? To an ad network? To the company that makes him?

People who love their pets often cite companionship as one of the pleasures of the relationship. When you are out with your dog (or home with your cat) you are not alone. Nor are you alone when you are with your robot companion. But it is not clear who it is that you are with.

Notes

1 In this chapter, I am assuming that robots are not sentient. They do not have a conscious experience of self. They are, however, capable of imitating being sentient and having emotions. Many roboticists and others have argued that machines are capable, someday, of becoming conscious. I do not wish to argue that point here, and it is mostly agreed that today's (and the near future's) robots, though they may seem remarkably aware, do not have what we would consider feelings, self-awareness, consciousness, and so on.
2 See (Archer, 1996) on why we love pets.

References

AFP-JIJI. (2015). An afterlife for man's best robot friend? *Japan Times*. www.japantimes.co.jp/news/2015/02/25/business/tech/afterlife-mans-best-robot-friend/#.WQnDenm1suU

Archer, J. (1996). Why do people love their pets? *Evolution and Human Behavior*, 18(4), 237–259. doi: 10.1016/s0162–3095(99)80001–4

Arkin, R. C., Fujita, M., Takagi, T., & Hasegawa, R. (2003). An ethological and emotional basis for human–robot interaction. *Robotics and Autonomous Systems*, 42(3–4), 191–201. doi: 10.1016/S0921–8890(02)00375–5

Breazeal, C. (2011). Social robots for health applications. Paper presented at the 2011 Annual International Conference of the IEEE Engineering in Medicine and Biology Society, August 30–September 3.

Calo, R. (2009). People can be so fake: A new dimension to privacy and technology scholarship. *Penn St. L. Rev.*, 114(3), 809–855.

Cullen, H., Kanai, R., Bahrami, B., & Rees, G. (2014). Individual differences in anthropomorphic attributions and human brain structure. *Social Cognitive and Affective Neuroscience*, 9(9), 1276–1280. doi: 10.1093/scan/nst109

Darling, K. (2017). "Who's Johnny?" Anthropomorphic framing in human-robot interaction, integration, and policy. In G. B. P. Lin, K. Abney, & R. Jenkins (Eds.), *Robot ethics 2.0*. Oxford: Oxford University Press.

Donath, J. (2004). Artificial pets: Simple behaviors elicit complex attachments. In M. Bekoff (Ed.), *The encyclopedia of animal behavior*, Vol. 3 (pp. 955–957). Westport, CT: Greenwood Press.

Donath, J. (2007). 1964 Ford Falcon. In S. Turkle (Ed.), *Evocative objects* (pp. 153–161). Cambridge, MA: MIT Press.

Donath, J. (2011). Pamphlets, paintings, and programs: Faithful reproduction and untidy generativity in the physical and digital domains. In T. Bartscherer & R. Coover (Eds.), *Switching codes: Thinking through digital technology in the humanities and the arts*. Chicago: University of Chicago Press.

Eisenberg, N., & Miller, P. A. (1987). The relation of empathy to prosocial and related behaviors. *Psychological Bulletin*, 101(1), 91–119.

Friedman, B., Kahn, P. H., Jr., & Hagman, J. (2003). Hardware companions?: What online AIBO discussion forums reveal about the human–robotic relationship. Paper presented at the Proceedings of the SIGCHI conference on Human Factors in Computing Systems, Ft. Lauderdale, Florida, USA

Fujita, M. (2004). On activating human communications with pet-type robot AIBO. *Proceedings of the IEEE*, 92(11), 1804–1813.

Fujita, M., & Kitano, H. (1998). Development of an autonomous quadruped robot for robot entertainment. Paper presented at the First International Conference on Autonomous Agents, February.

Ham, J., & Midden, C. J. H. (2014). A persuasive robot to stimulate energy conservation: The influence of positive and negative social feedback and task similarity on energy-consumption behavior. *International Journal of Social Robotics*, 6(2), 163–171. doi: 10.1007/s12369–013–0205-z

Ham, J., Bokhorst, R., Cuijpers, R., van der Pol, D., & Cabibihan, J.-J. (2011). Making robots persuasive: The influence of combining persuasive strategies (gazing and gestures) by a storytelling robot on its persuasive power. Paper presented at the Social Robotics, Third International Conference, ICSR, Amsterdam, The Netherlands, November 24–25.

Hartzog, W. (2015). Unfair and deceptive robots. *Maryland Law Review*, 74(785).

Horowitz, A., & Hecht, J. (2014). Looking at dogs: Moving from anthropocentrism to canid umwelt. In *Domestic dog cognition and behavior* (pp. 201–219). Berlin: Springer.

Johnstone, B. (2000). California dreamin' Sony style. *Technology Review*, 103, 72–79.

Kaplan, F. (2000). Free creatures: The role of uselessness in the design of artificial pets. Paper presented at the 1st Edutainement workshop.

Kim, E. (2016). Amazon's Echo and Alexa could add $11 billion in revenue by 2020. *Business Insider*, Sepember 23.

Kondo, M. (2014). *The life-changing magic of tidying up: The Japanese art of decluttering and organizing*. Berkeley, CA: Ten Speed Press.

Kopytoff, I. (1986). The cultural biography of things: Commoditization as process. In A. Appadurai (Ed.), *The social life of things*. Cambridge: Cambridge University Press.

Kretschmer, A. (2000). Mortuary rites for inanimate objects: The case of Hari Kuyō. *Japanese Journal of Religious Studies*, 27(3–4): 379–404.

Marinelli, L., Adamelli, S., Normando, S., & Bono, G. (2007). Quality of life of the pet dog: Influence of owner and dog's characteristics. *Applied Animal Behaviour Science*, 108(1–2), 143–156. doi: 10.1016/j.applanim.2006.11.018

Melson, G. F., Kahn, Jr., P. H., Beck, A., Friedman, B., Roberts, T., Garrett, E., & Gill, B. T. (2009). Children's behavior toward and understanding of robotic and living dogs. *Journal of Applied Developmental Psychology*, 30(2), 92–102. doi: 10.1016/j.appdev.2008.10.011

Mori, M. (2012). The uncanny valley [From the field], trans. K. F. Kageki & N. Kageki. *IEEE Robotics & Automation Magazine*, 19(2), 98–100. doi: 10.1109/mra.2012.2192811

Niculescu, A., van Dijk, B., Nijholt, A., Li, H., & See, S. L. (2013). Making social robots more attractive: The effects of voice pitch, humor and empathy. *International Journal of Social Robotics*, 5(2), 171–191.

Pogue, D. (2001). Looking at Aibo, the Robot Dog. *New York Times*, January 25. www.nytimes.com/2001/01/25/technology/looking-at-aibo-the-robot-dog.html

Rots, A. P. (2015). Sacred forests, sacred nation: The Shinto Environmentalist paradigm and the rediscovery of "Chinju no Mori." *Japanese Journal of Religious Studies*, 42(2), 205–233.

Rowling, J.K. (1999). *Harry Potter and the Chamber of Secrets*. New York: Scholastic Inc.

Siegel, M., Breazeal, C., & Norton, M. I. (2009). Persuasive robotics: The influence of robot gender on human behavior. Paper presented at the Intelligent Robots and Systems 2009 IEEE/RSJ International Conference on Intelligent Robots and Systems.

Sproull, L., Subramani, R., Walker, J., Kiesler, S., & Waters, K. (1996). When the interface is a face. *Human Computer Interaction*, 11, 97–124. doi: 10.1207/s15327051hci1102_1

Tam, K.-P., Lee, S.-L., & Chao, M. M. (2013). Saving Mr. Nature: Anthropomorphism enhances connectedness to and protectiveness toward nature. *Journal of Experimental Social Psychology*, 49(3), 514–521. doi: 10.1016/j.jesp.2013.02.001

Turkle, S. (2007). Authenticity in the age of digital companions. *Interaction Studies*, 8(3), 501–517.

Turkle, S. (2011). *Evocative objects: Things we think with*. Cambridge, MA: MIT Press.

Waytz, A., Cacioppo, J., & Epley, N. (2014). Who SEES HUMAN? The stability and importance of individual differences in anthropomorphism. *Perspectives on Psychological Science: A Journal of the Association for Psychological Science*, 5(3), 219–232. doi: 10.1177/1745691610369336

Weizenbaum, J. (1966). ELIZA: A computer program for the study of natural language communication between man and machine. *Communications of the ACM*, 9(1), 36–45.

Weizenbaum, J. (1976). *Computer power and human reason*. San Francisco: W. H. Freeman.

Weizenbaum, J. (1977). Computers as "Therapists" [letter to the editor]. *Science*, 198(4315), 354–354.

Wu, T. (2016). *The attention merchants: The epic scramble to get inside our heads*. New York: Knopf.

3

SELF IN THE LOOP: BITS, PATTERNS, AND PATHWAYS IN THE QUANTIFIED SELF

Natasha Dow Schüll

Which is the kind of being to which we aspire?

—Michel Foucault (1984)

We lack both the physical and the mental apparatus to take stock of ourselves. We need help from machines.

—Gary Wolf (2010)

In 1990, just as digital information and communication technologies were coming into widespread use, the French philosopher Gilles Deleuze (1992) suggested that the architectural enclosures, institutional arrangements, and postural rules of disciplinary societies were giving way to the networked technologies of "control societies," involving continuous coding, assessment, and modulation. The latter scenario bears an uncanny resemblance to the tracking-intensive world of today, in which the bodies, movements, and choices of citizens and consumers are ever more seamlessly monitored and mined by governments and corporations. The capacity to gather, store, and analyze individuals' physiological, behavioral, and geolocational data has come to affect a wide array of domains, from policy making to policing, corporate marketing to healthcare, entertainment to education.

Scholars working in the emergent field of critical data studies argue that data-tracking has come to "permeate and exert power on all manner of forms of life" in societies that are robustly digitally networked (Iliadis & Russo, 2016, p. 2). Some emphasize the "asymmetric relations between those who collect, store, and mine large quantities of data and those whom data collection targets" (Andrejevic, 2014, p. 1673; see also Beer, 2009; boyd & Crawford, 2012; van Dijck, 2014). Individuals perform unpaid and invisible "digital labor" in their role as data sources (Lupton, 2016, p. 118; Till, 2014); their data streams become "biocapital" to harness and exploit (Rabinow & Rose, 2006, p. 203). Selves are "sliced and diced into decon-textualized parts, and bought and sold," write anthropologists Neff and Nafus (2016, p. 62). Others focus on the threats that data-analytical *algorithms* pose to human agency, as "operations, decisions and choices previously left to humans are increasingly delegated to analytical algorithms, which may advise, if not decide, about how data should be interpreted and what actions should be taken as a result" (Mittelstadt et al., 2016, p. 9; Beer, 2009; Lash, 2007; Mackenzie, 2005).

Yet even as heated academic and public discussion unfolds around the dangers of data technologies, people have invited sensors to gather information about them through mobile apps and networked devices, convert this information into electrical signals, and run it through algorithms programmed to reveal insights and, sometimes, inform interventions into their future behavior. The contemporary world is characterized by "an intimacy of surveillance encompassing patterns of data generation we impose on ourselves," writes anthropologist Joshua Berson (2015, p. 40). As prescient as Deleuze's vision of the future was, Berson notes that the philosopher did not anticipate the degree to which the tracking and coding of bodies and acts would be drawn into the project of self-formation and self-care.

Michel Foucault (1988, p. 18) distinguished between technologies of power, "which determine the conduct of individuals and submit them to certain ends or domination, an objectivizing of the subject," and technologies of the self—through which individuals perform "operations on their own bodies and souls, thoughts, conduct, and way of being, so as to transform themselves in order to attain a certain state of happiness, purity, wisdom, perfection, or immortality." The latter take a literal, material form in the assemblages of sensors, analytical algorithms, and data visualizations that constitute contemporary self-tracking practices. This chapter brackets the growing literature on how data tracking serves as a technology of power to explore how it can also serve as a technology of the self—as a means, a medium, and a cipher for human experience, self-understanding, and sense of agency. Such an inquiry is a worthwhile endeavor not only in itself but also because it allows a richer understanding of datafication and its dynamics, and a more effective critique of its asymmetries and discontents. My point of entry for this exploration is the Quantified Self (QS) community.

Quantify Thyself

Nearly a decade ago in the San Francisco Bay Area, small groups of technologically savvy, existentially inquisitive individuals began to gather and reflect on what they might learn from data-gathering devices and analytical software about the mundane mysteries, dynamics, and challenges of their day-to-day lives—drug side-effects, sleep disorders, the association between diet and productivity. One at a time, they would "show and tell" their experiments in self-data, delivering 10-minute presentations scripted to answer a trio of framing questions: *What did you do? How did you do it? What did you learn?* After sharing their experiences, speakers would entertain questions and solicit feedback from those in attendance.

The group was anointed Quantified Self (QS) and, evoking the Delphic maxim "know thyself," given the tagline "self-knowledge through numbers" by co-founders Gary Wolf and Kevin Kelly, both former editors of *Wired* magazine. Through social media like meetup.com, QS quickly established a presence in major urban areas across North America and Europe, drawing in newcomers through a website featuring videos of members' presentations, a message board where people could discuss tracking tools, and links to local meet-ups. Today Wolf describes the group as "a loosely organized affiliation of self-trackers and toolmakers who meet regularly to talk about what we are learning from our own data" (2016, p. 67).

QS gained national prominence in the United States in April 2010, when a long-form essay by Wolf, "The data-driven life," ran as the lead article in the *New York Times Sunday Magazine*, a human figure collaged from graph paper, calipers, and folding rulers appearing on the cover. The article proposed that data could serve not only as a means of inspecting others' lives (as an actuary, policy maker, or welfare officer might) but also as a new kind of digital mirror in which to see and learn new things about ourselves. "Humans have blind spots in our field of vision and gaps in our stream of attention," wrote Wolf, "We are forced to steer by guesswork. We go with our gut. That is, some of us do. Others use data" (2010). In heart-rate spikes or mood dips charted over time, individuals could better grasp how they were affected by seemingly trivial habits or circumstances than by relying on expert advice, guesswork, or even intuition. "If you want to replace the vagaries of intuition with something more reliable, you first need to gather data," insisted Wolf. "Once you know the facts, you can live by them." In this do-it-yourself formula for self-care, data-intensive technology such as automated sensors, enumerative metrics, and statistical correlation were presented as tools for the good life.

Most who posted in the comments section for the story expressed disdain for the intensively tracked and monitored life that Wolf had prescribed, diagnosing it as a "loss of human-ness." A woman from New Jersey asked: "When do we reach the point that all we're doing is logging data for a life that's not being lived?" A reader from Kansas wondered what of lived experience we might miss by dwelling on "how many licks it takes to eat a lollipop" while another from Philadelphia wrote that "we are not machines and no amount of data will make us so—and no amount of data will give us all the answers to the bigger mysteries." The general response was that an excessive emphasis on that which can be measured degrades existence, rendering the unquantifiable stuff of life as so much noise to be filtered out.

A similar sentiment ran through the stream of one-off journalistic profiles of extreme self-trackers that appeared between 2010 and 2013 in the pages of *Forbes, Vanity Fair,* and even *Wired* itself, accounts that typically portrayed their subjects as caricatures of technological boosterism and American individualism (e.g. Hesse, 2008; Morga, 2011; Bhatt, 2013). Relying largely on these pieces, the cultural critic Evgeny Morozov (2013) launched an acerbic attack on the quantified-self community, alleging that its abandonment of narrative reflexivity in favor of soulless numerics was both dehumanizing and politically troubling. In a similar vein, social scientists have been quick to pin self-trackers as exemplary figures of contemporary biopolitics and governmentality, responsibly managing and optimizing their lives as cogs in a neoliberal wheel (Lupton, 2016; Ajana, 2017; Depper & Howe, 2017; Rich & Miah, 2017; Oxlund, 2012).

Yet an emerging body of ethnographic research has begun to pull the curtain back on a more nuanced reality, challenging the idea that quantified selves are necessarily existentially impoverished, depoliticized, or exploited. "QS is one of the few places where the question of why data matters is asked in ways that go beyond advertising or controlling the behaviors of others," write Nafus and Sherman (2014, p. 1788). As self-trackers readily acknowledge, quantification "rarely produces a definitive truth, a one-to-one representation of one's life or one's identity" (Sharon, 2017); instead, it involves a "situated objectivity" (Pantzar & Ruckenstein, 2017) in which certain prior experiences, understandings, and shared expectations come to matter. Sherman (2016) has described self-tracking as an aesthetic practice in which bits of the self, extracted and abstracted, become material for differently seeing and experiencing the self. Data is a kind of "transducer" that preserves only some qualities of the thing being measured such that "there is much room for people to maneuver in the imperfect translation" (Neff & Nafus, 2016, p. 25). Looking at personal data charts and visualizations can trigger critical reflection and raise new questions to pursue; the data does not displace or freeze but rather enhances and enlivens self-narratives (Ruckenstein, 2014, van den Eede). Self-quantification "sets up a laboratory of the self" in which "devices and data contribute to new ways of seeing the self and shaping self-understanding and self-expression" (Kristensen & Ruckenstein, 2018, p. 2). Such research approaches self-tracking as a form of open-ended experimentation in datafied subjectivity that is potentially noncompliant, creative, and reflexive—with as yet undetermined individual and collective possibilities.

This experimentation comes to the fore in the following scenes and conversations, which unfolded among participants in a two-day Quantified Self meeting in 2013.[1]

Seeing the Signal

After the 400-odd conference attendees had settled in their seats in the airy main hall of an Amsterdam hotel for a weekend of presentations and discussions, Gary Wolf took the stage to open the proceedings with a question: What exactly is a *quantified* self? Clearly, "quantification" involved collecting and computing data about ourselves, but "self," he ventured, was a more ambiguous term. How to understand the self in quantified self? What happens to the self when we quantify it—when "computing comes all the way in"?

Robin Barooah, a British technology designer now working in Silicon Valley, offered his answer to that question in the first show-and-tell session following Wolf's address. Wearing his signature fleece jumpsuit, he used a mixture of data visualization and personal backstory to share how he had measured his mood. Less fleeting than emotion but not as entrenched as temperament, "mood is mysterious," he noted. Robin had been drawn to a quantified-self approach to mood because he thought it could help him find non-intuitive, non-obvious connections between his life circumstances, daily habits, and mood. In 2008, a year he described as the most painful of his adult life, finding these connections was a matter of necessity rather than curiosity or self-experimentation: "I had to start examining my life and work out what to do."

"This isn't statistical analysis," he reassured the audience as he gestured at the timeline of data projected on the large screen behind him, mindful that not all conference attendees were versed in such a technique. The timeline spanned four years and plotted two variables whose relationship to his mood he was curious to learn: above the line, in blue, appeared the amount of time he had meditated daily, tracked with a timer;[2] below the line, in red, appeared the number of entries he had made each day in an online calendar set up to track his mood (see Figure 3.1). The choice to plot *how many* entries he was putting in the journal rather than some measure of their semantic content—a rating of the relative turmoil or

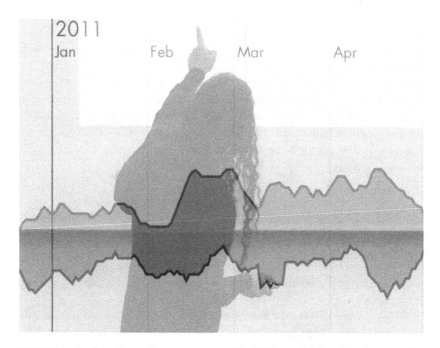

FIGURE 3.1 Robin Barooah on stage at QS 2013, explaining his data timeline (screenshot from video of presentation, available online at https://vimeo.com/66928697).

calmness expressed in his words, for instance—was deliberate: it was a way to measure the practice of journaling and see what it might reveal about his mood. What he was surprised to see when he finally (years later) plotted his data on minutes meditated and entries written was the uncanny correspondence between those two variables. "The coupling between the two lines is very clear," he told us. On any given day, more meditation was mirrored by more journaling, and vice versa. The tight correspondence gave his timeline the look of a Rorschach inkblot turned on its side, its top half in blue and its bottom half in red.

Robin drew the audience's attention to particularly volatile moments in the moving averages along the timeline: "this dip is where I was flying a lot"; "this trough with no color at all is a time of crippling anxiety and depression." Travel and major life events, including the death of his father, decoupled otherwise cor-related routines or lessened their symmetry. He pointed out a spot on the timeline with a large glacier of red activity beneath it and explained that a new psychop-harmacological regimen had spurred a period of intensive journaling. "A massive amount of narrative began to unfold at that time," Robin remembered. The color red falls off his timeline entirely in November of 2012, at which point the inten-sity of his anxiety had for the most part resolved and he no longer felt "the same impulse" to write in his journal: "it was a support that took effort, that involved ruminating on how I was feeling, and it basically felt better not to do it anymore."

Contemplating the visualization, Robin reviewed what he had learned. The graphical overlaying of disparate tracking routines, though not a direct representa-tion of his mood, was profoundly revelatory: "It's a kind of signal, one I hadn't seen before; it reflects my activation level, my energy level, my ability to engage with the world… it's there as an envelope around my whole life, affecting everything I'm doing." Being an engineer, Robin uses the word "signal" to describe information transmission—in this case, the conveyance of energies that powerfully affect his lived experience into a form he can perceive and assimilate. The signal, communi-cated in a a shifting silhouette of numeric values, is "beyond words … it's this very deep *thing* that ultimately becomes thinking and becomes action."

Discussing the Data

The next day, a breakout session on the theme of data and identity commenced with a set of questions posed by its convener, Sara Watson, a self-tracker and tech writer who had recently completed a master's (thesis) on QS practices 2013: *What does it mean to have data about myself—a digital, numerical, binary representation of myself? And what is my relationship to that data—what does it mean to be a human interacting with a digital binary thing that is data?*

Whitney, a self-tracker who regularly contributed thought-provoking pieces to the blog *Cyborgology*, suggested that data served as material for self-narratives: "we make stories about ourselves from the data, to make sense of our lives." Some in the room pushed back, wanting to preserve the facticity of data as expressing an objec-tive truth: data was not some "made up" story; if anything, QS *denarrativized* the self.

Joshua, a bearded venture capitalist in his early thirties from California, elaborated on this idea: "The self can be overwhelming as an integrated, whole thing. By doing QS, you can *disaggregate* various aspects of self, work on just those, maybe let them go, put them back in … It takes an incredible burden off you when you can take these small slices out and say, *all that other stuff is complicated, let's just look at this.*" Robin interjected to reinforce this point: "Tracking isn't additive—it's *subtractive*: you work on some question about yourself in relation to this machine-produced thing and you know that *it will stop*; afterward, you're left with a narrower range of attributions you can make about your behavior or your feelings; you have eliminated uncertainty and gained a kind of liberation—you can move on with your life, with a new perspective." If this extractive, bitifying process was a form of self-narration, Joshua concluded, then we should call it "quantitative autobiography."

Joerg, a German activist whose background in business and philosophy complemented his pursuit of a data-based ethics in the corporate world, further specified the term "narrative" as it pertained to self-quantification: "Numeric expressions of ourselves are inherently *syntactic*, not *semantic*." The power of self-data lay in the relational grammar that emerged across its data points—not in the authorial intentions of "transcendent phenomenal selves" storying themselves forth. His position at once echoed and countered Morozov's criticism: yes, self-quantification departed from traditional humanist modes of narrative—but that did not make it *dehumanizing*; rather, it was vital, enlivening.

An American anthropologist employed at a leading technology firm suggested that art, rather than narrative, might be a better metaphor to describe what selves do with their data. "Maybe tracking is like *sketching* yourself," mused another participant in the session. "You have to fill in the details, it's a kind of self-portrait, an art." Robin, from his seat along the side wall, nodded in agreement. He remarked that he had once characterized his tracking as a kind of "digital mirror" but now felt the metaphor was inaccurate, "because mirrors represent a whole, projected image—which is not what we get from our data bits." Returning to the earlier point he and Joshua had made, he suggested that the value of data points tracked in time is the *narrowness* of the representation they provide: "Data is really just numbers, symbols—it doesn't reflect back something that already exists in the world as a mirror does; instead it shows us a model of some limited, extracted aspect of ourselves." Robin had come to prefer the metaphor of self-portraiture: "What we're doing when we track and plot our data is focusing in on one part of our lives and slowly building up that portrait as we collect data on it."

Sara, our moderator, pressed the group to further specify the metaphor: If not photorealistic, was the portrait expressionist? Impressionistic? Pixelated? "I think it would have to be an algorithmic mosaic, with shifting composition, color, and patterns, an ever-changing portrait." Robin suggested. "But *in what way* does it change?" asked a fellow tracker, voicing some ambivalence over his relationship to his data. "I only look at bits and pieces of myself because it's all I can handle. If it's a portrait, then it's a portrait with really bad lighting … Isn't the point, ultimately, to shine a brighter light on ourselves? Does the portrait ever gain fuller resolution, become more solid, more like a true mirror?"

Joerg posed the question as a tension between self-making and self-*un*making: "If you start breaking yourself down piece by piece, it could lead to non-self, disaggregation, seeing ourselves as a big stream of data … Or can it, somehow, make us feel more *solid* as selves in the world?" Robin ventured that there was no contradiction between self-making and unmaking:

> I think they're consistent views really. If self-quantification, breaking ourselves down into bits, enables us to create new experiences of ourselves, then those experiences are gateways to *new degrees of freedom* in how to act.

The kind of portraiture at stake in the Quantified Self, he suggested, "allows you to imagine new types of self and move in new directions; you are no longer trapped in a limited set of pathways." Self-tracking, it seemed by the end of the discussion, was a means of liberation not only from the impasses of uncertainty but those of certainty as well.

Time-Series Selves

Eric Boyd, a mechanical engineer known in the QS community for designing pendants that flash in time with wearers' heartbeats and vocal cadence,[3] delivered a Show and Tell on the second day of the conference, sharing insights into his "daily rhythms" gleaned from his (since-discontinued) Nike Fuelband, a rubberized accelerometer worn on the wrist. He admitted being drawn to the "geeky bling factor" of the consumer gadget and its colorful, sequentially blinking lights, but was otherwise unimpressed. "The graphs on the app are pretty but mostly useless; you can't even tell what time of day things happened. It was super frustrating how non-visible my activity was." The analytic features provided for users obfuscated their activity as so many inscrutable "fuel points"—a measure of activity proprietary to Nike.

Wanting to examine his daily patterns more closely, Eric interfaced with the Fuelband's object-oriented programming language[4] to feed the raw values from the accelerometer into a spreadsheet, rendering one cell for every minute of the day and one column for every day of the month: "1440 rows by 30 columns— that's a lot of data showing what I was doing when." He was able to see when he woke up at night to visit the bathroom, and that his usual brisk pace became slower when walking with his girlfriend. Her walking speed was something of an issue in their relationship, he admitted, "and it helped to see that it was actually only 30 percent slower."

Eric's self-tracking concerned habits and life rhythms more prosaic than the depressive troughs and intense peaks at stake in Robin's, but the core features of self-inquiry were the same: a question; an investigation that apportioned significant epistemological authority to data and its technologies; visual reveals; unexpected discoveries. "The reason you begin tracking your data is that you have some uncertainty about yourself that you believe the data can illuminate," Eric told me. "It's about introspection, reflection, seeing patterns, and arriving at realizations about who you are and *how you might change*." His "introspection," like Robin's, commences not with a turn inward but a turn outward to the streaming data of a device: an extraction of information, a quantification, a visualization.

"When we quantify ourselves, there isn't the imperative to see through our daily existence into a truth buried at a deeper level," Wolf (2010) wrote. "Quantified self is not a linguistic exploration like psychoanalysis," echoes Eric, "it's a digital exploration, and the stuff you're exploring is made up of many little bits and moments." Bits and moments, accumulating into habits, rhythms, and tendencies, are the "stuff" of the self—their effects cannot be known through the semantic twists of language unfolding in the moment of a human encounter (as in therapeutic transference); rather, they must pass through an array of sensors, numeric representations, and algorithmic processes that detect their relational patterns over time (Nafus, 2016, p. xviii; Sherman, 2016; Day & Lury, 2016). "You may not gain any knowledge in a week or even a month," says Eric, "but over time you might see something significant about yourself; you need a view that's longer than whatever moment you're in."

A few years ago, out of concern for climate change, Eric decided to track his driving habits. He knew how many miles he was putting on his vehicle but was not certain which of his routines—going to work, going on road trips, going out socializing—was most significant. "So I tracked every single car trip for around three months and then I put it all into an Excel spreadsheet, with different destinations into categories to see what was driving my miles." He learned that his daily trips to work, only a few kilometers away, were the major contributor to his mileage. "My work was only around 3.5 km, so I hadn't thought it would be significant—but it added up because I would do it around two times a day, and often I would have to circle around the block to find parking. So the *accretion of those little trips* added up to at least as much as the road trips and the socializing."

By engaging data and its technologies to assist in his self-inquiry, Eric does not lose agency so much as he finds a new kind of agency. "In our physical world," he explains, "our powers only extend a few meters—but in the *temporal* dimension we're extremely effective, we're actually going to live a billion moments or something like that. The trouble for us is that *it's difficult for us to see the amount of power we have in time because our sense of time is so limited*; we go through life one minute a time." Data tracking and time-series analysis "give a longer view of our power in time" by showing how our habits—"the things we're doing over and over"— add up to affect our lives in positive and negative ways. As Wolf (2010) writes, "without good time calibration it is much harder to see the consequences of your actions." Through tracking, Eric has come to regard himself as a "time-series self," one whose truth and consequences are not fixed but made of small actions over which he has some measure of control; like Robin, he finds this vantage liberating and empowering.

Selves in the Loop

Over the past five years, scholars have reached for creative vocabulary to describe how the intensive datafication of life in Western liberal societies is altering self-hood. Some apply Deleuze's prescient notion of the *dividual*[5] as an archive of traits, habits, and preferences that can be systematically extracted and marketed to, reshuffled, and compared to those of others. Harcourt (2015) gives the term a twist, proposing that *duodividual* better describes "the object of the algorithmic data-mining quest of the digital age"—which, he argues, is not so much to divide an individual into parts but to find its digital "match" or "twin." Many adopt the term *data double* (Haggerty and Ericson, 2000) to name this digital doppel-gänger, while for Horning (2012) *data self* best describes the virtual versions of ourselves that arise through social media. "Disembodied exhaust gives rise to a *data-proxy*, an abstracted figure created from the amalgamation of data traces," writes Smith (2016, p. 110). Greenfield (2016, p. 133) discusses the "pixelated person" as "a subject ever divided into finer granularity, but also whose partial datasets can be joined with others." Switching from the register of data to that of the algorithm, Cheney-Lippold (2011, p. 165) describes *algorithmic identities* as "categories of identity [that] are being inferred upon individuals based on their web use" while Pasquale (2015, p. 39) argues that we are treated as *algorithmic selves* or "set[s] of data points subject to pattern recognition engines." For all the subtle nuances and asymmetries this array of neologisms captures about the processes of fragmentation, amalgamation, and aggregation through which selves are made objects and subjects of power in a digitally networked world, they are less helpful when it comes to grasping how selves inhabit, experience, reflect on, and act in their datafied lives.

As we have seen, in archived sequences and sums of bitified life, quantified selfers seek to bring to awareness the patterns and rhythms that define their existence and that might, without digital tools, remain uncertain forces below the threshold of perception.[6] In this sense, they follow the recommendation of the media scholar Mark Hansen (2015, p. 38) that we "forge connections" with machinic capacities and microtemporal processes, even as they evade the "grasp of our conscious reflec-tion and sense perception." Technical data-gathering and analysis, he emphasizes (p. 196), can be used "not solely to anticipate our tendencies and susceptibilities for purposes of manipulation and exploitation, but also to inform us about these tendencies and susceptibilities and let us act on and in virtue of them."

While admitting machinic forms of intelligence into human ways of defining, categorizing, and knowing life risks the loss of human autonomy, it also car-ries the possibility for new human agencies (Kristensen & Ruckenstein, 2018; Kennedy et al., 2015; Neff & Nagy, this volume). The science and technology pundit Melanie Swan (2013, p. 95) proposes that big-data epistemologies, transposed to the scale of the individual, afford "a sort of fourth-person perspective" on the self and, ultimately, a new kind of truth—one that is "not possible with ordinary senses." This truth does not correspond to a classical phenomenological self grounded in time and space but to a "database self" (Schüll, 2016a, 2016b) that extends over time. "You set up this kind of external person or version of yourself, an avatar or companion—or something," said a tracker during Watson's breakout session in Amsterdam, recalling Foucault's (1998, p. 211) characterization of self-care as "establishing a relationship of oneself with oneself." "I had arrived at a place where it was necessary to start relating to myself," a QS member told two anthropologists (Kristensen & Ruckenstein, 2018, p. 9). "It's all about finding a direction, gaining awareness, and then arriving at a 'felt' understanding of oneself," said another (p. 8).

As Robin told us earlier, archives of personal data bits, continuously recomposed, can be "gateways to *new degrees of freedom* in how to act." The kind of freedom he invokes here is not the freedom of autonomy or self-mastery but, rather, as Colin Koopman (2016, p. 45) has characterized the philosophy and life of William James, "freedom amid uncertainty as the work of self-transformation," with the self understood to be "the activity of reflexive recomposition" (p. 47). In the case at hand, digital tracking tools and their data can become part of the loop of reflexive recomposition—expressed as a practice of "quantitative autobiography," "continuous self-portraiture of shifting composition," and the computational-graphic emplotment of signals "beyond words." These emergent conceptions well articulate this chapter's framing notion of "self in the loop" (itself a reference to the AI term "human-in-the-loop," in which human operators or users can alter the outcome of a computational event or process) and stand as important accompaniments to the scholarly neologisms that seek to describe the fragmented, alienated, and exploited selves of the datafied world.

For James, self-transformational ethics entailed "instigating alternatives, provoking differentia, becoming undisciplined and even undisciplinable" (Koopman 2016, p. 43). Likewise, for self-trackers metrics can serve for "detouring from prescribed courses, exploring limits, and defying rules" (Sanders 2017, 21). Nafus and Sherman (2014, p. 1785) write of self-tracking as a form of "soft resistance" that is "always necessarily partial, firmly rooted in many of the same logics that shape the categories they seek to escape." Rather than dismiss trackers of self-data—as life-avoiding and robotically inclined; as victims of data capitalism and its surveillance apparatus; or as symptomatic figures of neoliberal subjectivity and its self-mastering, entrepreneurial ethos—we might regard them as pioneers in the art of living with and through data. Inviting digital tools and epistemologies to partake in their self-transformational ethics, quantified selfers gain new methods for apprehending, knowing, and inhabiting their lives—and, potentially, for resisting, repurposing, and rendering uncertain the normative proxies, behavioral categories, and governing logics that would seek to drive their conduct down certain pathways.

Notes

1 This chapter draws on ethnographic research conducted between 2013 and 2017 at QS meet-ups in Boston and New York as well as three annual conferences. At these events I listened, observed, and spoke informally and formally with participants; once in San Francisco and twice in Amsterdam, I convened and led conference sessions. Where recordings are available (my own or posted online), I quote quantified selfers directly; in other cases I draw on fieldnotes.
2 Barooah went on to develop a meditation app called Equanimity, along with a flash-based timer for internet browsers, available at www.meditate.mx.
3 Boyd's company, Sensebridge, has designed the Heart Spark pendant, which flashes in time with one's heartbeat, while the Sound Spark flashes along with the cadence of one's voice; a compass anklet works at a haptic register, vibrating to augment one's sense of direction.
4 Typically, such a maneuver involves breaking a product's warranty and/or terms of service agreement, but in this case, Nike offered advanced users an API to extract the information.
5 It should be noted that a different conceptual trajectory for the term *dividual* exists in anthropology to describe forms of selfhood that are not based in Western dualisms (e.g. Strathern 2004) and that are constituted by social relations rather than discrete units.
6 Hong notes that in self-tracking "some measurements, like galvanic skin response, are absolutely beyond human access; others, like steps taken, are measured with a frequency and precision practically unavailable to human subjects" (2016, p. 15).

References

Ajana, B. (2017). Digital Health and the Biopolitics of the Quantified Self. *Digital Health* 3 (January): https://doi.org/10.1177/2055207616689509
Andrejevic, M. (2014). The big data divide. *International Journal of Communication,* 8, 1673–1689
Beer, D. (2009). Power through the algorithm? Participatory web cultures and the techno-logical unconscious. *New Media and Society,* 11, 985–1002
Berson, J. (2015). *Computable bodies: Instrumented life and the human somatic niche.* New York: Bloomsbury.
Bhatt, S. (2013). We're all narcissists now, and that's a good thing. *Fast Company,* September 27, 2013. https://www.fastcompany.com/3018382/were-all-narcissists-now-and-thats-a-good-thing
boyd, d, and Crawford, K. (2012). 'Critical questions for big data: provocations for a cultural, technological and scholarly phenomenon.' *Information, Communication, & Society,* 15, 662–79.
Cheney-Lippold, J. (2011). A new algorithmic identity: Soft biopolitics and the modulation of control. *Theory, Culture & Society,* 28(6), 164–181. doi: 10.1177/0263276411424420
Day, S., & Lury, C. (2016). Biosensing: Tracking persons. In D. Nafus (Ed.), *Quantified: Biosensing technologies in everyday life* (pp. 43–66). Cambridge, MA: MIT Press.
Deleuze, G. (1992). Postscript on the societies of control. *October,* 59, 3–7.
Depper, A., & Howe, P. D. (2017). Are we fit yet? English adolescent girls' experiences of health and fitness apps. *Health Sociology Review,* 26(1): 98–112. https://doi.org/10.1080/14461242.2016.1196599.
Foucault, M. (1984). 'On the Genealogy of Ethics: An Overview of Work in Progress,' *The Essential Works of Foucault 1954–1984,* Ed. Paul Rabinow, Trans. Robert Hurley and Others, New York: The New York Press.

Foucault, M. (1988). Technologies of the Self. In L. Martin, H. Gutman, & P. Hutton (Eds.), *Technologies of the self: A seminar with Michel Foucault.* Cambridge, MA: University of Massachusetts Press.

Foucault, M. (1998.). Self-writing. In *Ethics: Subjectivity and truth*, edited by Paul Rabinow, (pp. 207–22). New York: The New Press.

Greenfield, D. (2016). Deep data: Notes on the n of 1. In D. Nafus (Ed.), *Quantified: Biosensing technologies in everyday life* (pp. 123–146). Cambridge, MA: MIT Press.

Haggerty, K. D., & Ericson, R. V. (2000). The surveillant assemblage. *The British Journal of Sociology*, 51(4), 605–622. doi: 10.1080/00071310020015280

Hansen, M. B. N. (2015). *Feed-forward: On the future of twenty-first-century media.* Chicago: University of Chicago Press.

Harcourt, B. E. (2015). *Exposed: Desire and disobedience in the digital age.* Cambridge, MA: Harvard University Press.

Hesse, Monica. 'Bytes of Life.' The Washington Post, September 9, 2008.

Hong, S. (2016). Data's intimacy: Machinic sensibility and the quantified self. *communication +1*: Vol. 5, Article 3. DOI: 10.7275/R5CF9N15

Horning, R. (2012). Notes on the "data self." *Marginal Utility* (blog). February 2. https://thenewinquiry.com/blog/dumb-bullshit/

Iliadis, A., & Russo, F. (2016). Critical data studies: An introduction. *Big Data Soc.* July–Dec., 1–7

Kennedy, H., Poell, T., & van Dijck, J. (2015). Introduction: data and agency. *Big Data Soc.* 2: https://doi.org/10.1177/2053951715621569

Koopman, C. (2016). 'Transforming the Self amidst the Challenges of Chance: William James on 'Our Undisciplinables.'' Diacritics, 44 (4), 40–65. https://doi.org/10.1353/dia.2016.0019.

Kristensen, D. B. & Ruckenstein, M. (2018). 'Co-evolving with self-tracking technologies.' *New Media & Society.* First Published February 21, 2018, 1–17. https://doi.org/10.1177/1461444818755650

Lash, S. (2007). Power after hegemony: cultural studies in mutation? *Theory, Culture & Society,* 24: 55–78

Lupton, D. (2016). *The quantified self: A sociology of self-tracking.* Cambridge: Polity Press.

Mackenzie A., (2005). The performativity of code: software and cultures of circulation. *Theory, Culture & Society,* 22: 71–92.

Mittelstadt, B. D., Allo, P., Taddeo, M., Wachter, S., & Floridi, L. (2016). The ethics of algorithms: mapping the debate. *Big Data Soc.*, 3. doi: 10.1177/2053951716679679

Morga, A. (2011). Do you measure up? *Fast Company,* April 5, 2011. https://www.fastcompany.com/1744571/do-you-measure.

Morozov, E. (2013). *To save everything, click here: The folly of technological solutionism.* New York: Public Affairs.

Nafus, D. (Ed.). (2016). *Quantified: Biosensing technologies in everyday life.* Cambridge, MA: MIT Press.

Nafus, D., & Sherman, J. (2014). This one does not go up to 11: The quantified self movement as an alternative big data practice. *International Journal of Communication*, 8(0), 1784–1794.

Neff, G., & Nafus, D. (2016). *Self-tracking.* Cambridge, MA: MIT Press.

Oxlund, B. (2012). 'Living by Numbers.' *Suomen Antropologi: Journal of the Finnish Anthropological Society*, 37 (3), 42–56

Pantzar, M., & Ruckenstein, M. (2017). Living the metrics: Self-tracking and situated objectivity. *Digital Health*, 3, January 1. doi: 10.1177/2055207617712590

Pasquale, F. (2015). The algorithmic self. *The Hedgehog Review.* www.iasc-culture.org/THR/THR_article_2015_Spring_Pasquale.php

Rabinow, P., & Rose, N. (2006). Biopower today. *BioSocieties*, 1, 195–217.

Rich, E., & Miah, A. (2017). Mobile, wearable and ingestible health technologies: Towards a critical research agenda. *Health Sociology Review*, 26(1): 84–97. https://doi.org/10.108 0/14461242.2016.1211486.

Ruckenstein, M. (2014). Visualized and interacted life: Personal analytics and engagements with data doubles. *Societies*, 4(1), 68–84. doi: 10.3390/soc4010068

Sanders R (2017) Self-tracking in the digital era: biopower, patriarchy, and the new biometric body projects. Body & Society 23(1): 36–63.

Schüll, N. (2016a). Abiding chance: Online poker and the software of self-discipline. *Public Culture*, 28(3), 563–592. doi: 10.1215/08992363–3511550

Schüll, N. (2016b). Data for life: Wearable technology and the design of self-care. *BioSocieties*, 11(3), 317–333. doi: 10.1057/biosoc.2015.47

Sharon, T. (2017). Self-tracking for health and the Quantified Self: Re-articulating autonomy, solidarity, and authenticity in an age of personalized healthcare. *Philosophy & Technology*, 30, 93–121.

Sherman, J. (2016). Data in the age of digital reproduction: Reading the quantified self through Walter Benjamin. In D. Nafus (Ed.), *Quantified: Biosensing technologies in everyday life* (pp. 27–42). Cambridge, MA: MIT Press.

Smith, G. J. D. (2016). Surveillance, data and embodiment: On the work of being watched. *Body & Society*, 22(2), 108–139. doi: 10.1177/1357034X15623622

Strathern, M. (2004). The whole person and its artifacts. *Annual Review of Anthropology*, 33(1), 1–19. doi: 10.1146/annurev.anthro.33.070203.143928

Swan, M. (2013). The Quantified Self: Fundamental disruption in big data science and biological discovery. *Big Data*, 1(2), 85–99. doi: 10.1089/big.2012.0002

Till, C. (2014). Exercise as labour: Quantified Self and the transformation of exercise into labour. *Societies*, 4, 446–462

van Dijck, J. (2014). Datafication, dataism and dataveillance: Big data between scientific paradigm and ideology. *Surveillance & Society*, 12(2), 197–208.

van den Eede, Y. (2015). Tracing the tracker: A postphenomenological inquiry into self-tracking technologies. In R. Rosenberger and P.-P. Verbeek (eds.), *Postphenomenological investigations: Essays on human–technology relations* (pp. 143–158). Lanham: Lexington Books.

Watson, S. (2013). Living with data: Personal data uses of the Quantified Self. MPhil dissertation. University of Oxford.

Wolf, G. (2010). The data-driven life. *New York Times*, April 28, sec. Magazine. www.nytimes.com/2010/05/02/magazine/02self-measurement-t.html

Wolf, G. (2016). The Quantified Self: Reverse engineering. In D. Nafus (Ed.), *Quantified: Biosensing technologies in everyday life* (pp. 67–72). Cambridge, MA: MIT Press.

4

POSTHUMAN FUTURES: CONNECTING/DISCONNECTING THE NETWORKED (MEDICAL) SELF

Laura Forlano

Networked medical devices, unlike communication technologies such as the internet, mobile phones, and social media, offer a unique perspective on the topic of the networked self in the context of human augmentics, artificial intelligence, and sentience. This chapter draws on feminist technoscience in order to understand how networked medical technologies allow us to reconsider what it means to be human, how we understand ourselves and how we create meaning in our lives. Through three autoethnographic vignettes, this piece reflects on the experience of living with/being/becoming part of these networked technologies as a form of posthuman subjectivity. Here, disconnection, rather than connection, emerges as a key theme that animates and makes visible the social and material experience of these technologies. These moments of failure and breakdown allow for new considerations around self-knowledge, agency, and actualization embedded in practices of repair and care.

Since the mid-2000s, the majority of my research and writing has engaged with the materialities of the digital. Amidst the constant claims of dematerialization in the last several decades, which continue to advance the argument that the digital, virtual, online, and networked realms are replacing the physical world, such research is vital for insuring that the aesthetics and politics around bodies, things, and environmental resources continue to matter and, as a result, are less easily dismissed. Specifically, I have been interested in the ways in which sometimes invisible sociotechnical systems and infrastructures reveal themselves through social and cultural practices. These practices are often made visible through the reconfiguration of bodies, representations, and things in time and space (i.e. the global distribution of work practices (Forlano & Mazé, forthcoming), the use of environmental resources, etc.). For example, in 2003, I spent hours watching Japanese teenagers using some of the first camera phones at Shibuya Crossing in Tokyo.

The following year, I sat day after day at Szoda, a cafe in Budapest located near the Great Synagogue, watching people use laptops on the free Wi-Fi network, which was one of the city's first. I hosted my own Wi-Fi hotspot out of my East Village apartment while simultaneously surveying over 600 people in three cities and interviewing over thirty people about the ways in which they used these technologies in 2006. That same year, I stood on the rooftops of cooperative housing blocks in what was formerly East Berlin in order to study community-run wireless networks (Forlano, 2009). I organized exhibitions and events in New York in order to better understand how the early versions of these networks and location-based social media could be used to bring people together in 2009 (Forlano, 2011; Townsend et al., 2011). I stood in the cold while testing a location-based application on a college campus in upstate New York in 2010 (Forlano, 2013). Over the past ten years, a growing body of literature has sought to make sense of these and related practices drawing on theories of digital materiality (Dourish, 2017; Forlano, 2015; Pink et al., 2016), which—simply put—is interested in explaining both the materiality of the digital and, at the same time, the digitality of the material.

This chapter engages with similar themes around digital materiality, however, unlike my previous projects, which had theories, methods, protocols, datasets, and many small black Moleskine notebooks covered with stickers and inscribed with notes and sketches, this project was unplanned, unwanted and, even, somewhat unusual. After being diagnosed with Type 1 diabetes in 2011 and adopting an insulin pump and continuous glucose monitor (CGM) in 2013, I began writing about my own experiences with these networked medical devices in a series of articles.

While this writing did not start out as research, my previous projects provided me with a theoretical filter through which to view this experience. It prepared me to pay attention to the specifics of the ways in which technologies—in this case the intimate infrastructures used for managing diabetes—are entangled with everyday lived experiences. In this way, I argue that becoming is a form of knowledge production that aligns with feminist technoscience. Through becoming, I participate in many research practices—calculation, measurement, calibration, testing, experimenting, data collection, data interpretation, and data sharing. I perform many kinds of labor to maintain and repair these networked medical devices. What makes this feminist technoscience is not these activities themselves (as they have much in common with other quantified self and DIY science communities). What make this feminist science is the integration of theory and practice around the body as an epistemic site, the attention to the relations between human and non-human actors, the understanding and acknowledgment that this is a partial perspective on the world, the awareness of the privilege and power that makes it possible for me to write this narrative at all, and the enactment of care that takes place when that knowledge is distributed and shared. Feminist technoscience attends to alternative sets of ethics, values, and responsibilities to the world(s) that it participates in.

By paying careful attention to and reflecting on the activities, rituals, and prac- tices that I engage in on an everyday basis, I am able to give an account of the embodied, lived experience of the networked (medical) self. These rituals include everyday tasks such as taking a shower, brushing one's teeth, and getting dressed, which must change to accommodate the new "features" that have been added to my body. For example, showering means removing the tubing by screwing off a circular cap and carefully scrubbing around the "port" so as not to get threads from my washcloth stuck on it. Brushing my teeth often means resting the pump on the counter while I am wrapped in a towel (without anywhere to attach it). When getting dressed, I choose my clothes in a constant negotiation with the pump. I stand in the closet, pump in one hand, and try on a series of outfits. I can wear nearly anything, but some combinations do not work as well as others. If I get creative, I find that I can rig the tubing out of a red wool sleeveless sweater and stick it into the left-hand pocket of a black blazer or attach it to the middle of my bra and cover it with a colorful scarf.

Some of these accounts are frustrating (like when the plastic clip on the pump's holster breaks while I am overseas), while others are more humorous (like when I get my tubing stuck on the knob of a kitchen drawer). Drawing on Haraway's cyborg (1991), I have created the figure of the *disabled cyborg* as a way of under- standing the ways in which networked medical devices offer to extend and repair the capabilities of my body while at the same time suffering their own break- downs (Forlano, 2016b). Here, I employed the concept of hacking because of its associations with cutting into, participating in (and breaking into) and coping. Also, because of the kind of comedic narrative that I wanted to tell in that piece, I was drawn to the ways in which hacking is linked to jokes, trickery, and hoaxes (Coleman, 2012).

Subsequently, I began to consider the ways in which I was participating in the maintenance, repair, and care of the devices that were ostensibly taking care of me. For example, there are a myriad of mundane, everyday failures that occur with regularity: the pump battery dies on the way to the office, tubes bend in half and prevent insulin from flowing to the body, the CGM or meter need to be recharged, sticky adhesive surfaces unstick, and readings are incorrect, to name just a few. Usually, these minor inconveniences can be handled quite easily in the course of one's everyday life. But, sometimes, multiple things fail at once or at times that are considerably awkward. In these situations, it is clear that the system, which otherwise often appears to be functioning quite well, is incredibly fragile and the risk of what could be a very life-threatening breakdown is real.

Drawing on Jackson's broken world thinking (Jackson, 2014), I develop a multi-scalar theory of *broken body thinking* that collapses the world onto the body and explores the disabled cyborg body as a multiple, posthuman, and more-than- human subjectivity (Forlano, 2016a, 2017a, 2017b), which suggested a differ- ent set of concerns, relations, contingencies, and processes. In my understanding of the posthuman, I include both considerations around the possibilities and constraints of technology as well as our relationship to ecologies and the natural environment. Rather than merely conceiving of myself as a human user of networked medical devices, I described my identity as a user-repairer, actively participating in practices of maintenance (Sample, 2016; Vinsel & Russell, 2016), repair and care (Mol, 2008; Rosner & Ames, 2014; Toombs et al., 2015). Drawing on Annemarie Mol's "logic of care" (in contrast to the "logic of choice" in the market economy), I describe the mutual, ongoing, collaborative, and collective work that is required every day. Here, it is useful to ask what notions around care might offer to continued conversations on science, innovation, and infrastructure with a specific focus on human augmentics, artificial intelligence, and sentience. But, what of the labor implications of this identity, as I take on more and more of the work that might have been done by institutions in the past. On the one hand, I have a great deal of autonomy in taking care of myself; on the other hand, I am doing the work that doctors or hospitals might have done in the past. If users must also be repairers, their work also needs to be compensated, valued, and recognized as such.

Drawing on feminist new materialism (Bennett, 2009; Braidotti, 2013) and discussions of intersectionality (Crenshaw, 1991), I considered the ways in which I participate in multiple social worlds and subjectivities (and thus related patterns of oppression and discrimination) related to gender, technology, disability, and difference. Feminist new materialism has also called my attention to the immense amount of medical waste—plastic, packaging, and needles—that is produced on a weekly basis in the everyday life of the disabled cyborg.

Building on this, more recently, I have engaged with Kafer's notion of crip time (Kafer, 2013) in order to explain the experience of difference faced during mundane everyday activities such as eating. In this work, I use Carey's ritual view of communications (Carey, 1988) to develop the idea of *data rituals*, which are the embodied processes required to manage and participate in everyday activities that require access to data about blood sugar (Forlano, forthcoming). I use the term *intimate infrastructures* to capture the ways in which our lives are bound up with sociotechnical systems and assemblages. Intimacy suggests alignment with feminist technoscience in its relationship to situated, contextual, embodied, affective, and partial knowledges (Haraway, 1988; Wajcman, 2000).

In thinking about posthuman futures, and, in particular, those introduced by human augmentics, artificial intelligence, and sentience, I return to Kafer's work on crip futures, which advocates for "futures that embrace disabled people, futures that imagine disability differently, futures that support multiple ways of being" (2013, p. 45). In particular, I aim to locate this narrative in emergent discussions around critical posthumanism, which accounts for race, gender, class, ability, sexuality, and geography (Weheliye, 2014). It is in this spirit, inspired in part by S. Lochlain Jain's *Malignant* (Jain, 2013), that I offer my account of living with networked medical devices.

Disconnection and Difference

In this chapter, I once again engage with many of the themes above with an attention to disconnection and the networked medical self, which contributes to growing interest in the quantified self in recent years (Dow Schüll, 2016; Kaziunas et al., 2017; Neff & Nafus, 2016). Disconnection, failure, and breakdown serve to make infrastructures more visible according to feminist STS theory (Star, 1999). When notions of infrastructures collapse onto the body, their breakdown and decay take on new and different meanings. Interoperability, standards, and intellectual property have different consequences. Such mundane personal examples serve to reveal the ways in which pervasive, simplistic, rational, universal, and positivistic notions of the human are complicated by intersectional experiences of difference. How does the use (and failure) of these technologies summon a particular kind of human subject and, as a result, how do I understand myself/ myselves? How do I experience difference and adjust the ways in which I engage with the world? What are my possibilities for the future?

When I first chose to adopt the insulin pump in 2013, I met with a representative from Medtronic who spent about an hour and a half talking to me, telling me anecdotes about how other clients—athletes running a marathon, teenagers going to the prom—had adjusted to using the pump. For example, he recounted the challenges in terms of eating, exercising, and socializing as well as the ways in which they were able to troubleshoot these problems.

One of these stories was about how to eat a pizza. Specifically, the pump has special settings that can better manage the dosage of insulin for eating foods high in saturated fat such as pizza, which tend to absorb into the body very slowly. As he convinced me of the benefits of the technology, he mentioned that it would never be "perfect." The pump is deeply entangled with the messy, material reality of the human body. Stress, hormones, exercise, and so many other biological, social, and environmental factors influence the body in many ways. So, despite the emphasis on routine for diabetics, you are never really eating the same thing twice under the same conditions. At least, under the guise of tech support, I now have someone to call for advice on how to eat a pizza.

We did not, however, discuss the many ways in which the devices themselves might falter, fail, or betray. I introduce some of these kinds of situations in the following vignettes. It is in reflecting on these minutes, hours and sometimes days of disconnection that these intimate infrastructures are revealed.

Brexit, Batteries, and Blood (Sugar)

In June 2016, just as the United Kingdom was in the middle of its own Brexit breakdown, I was attending an important design conference in Brighton. On the first day, during the first hour of a hands-on workshop on "Smart

Cities," the AAA battery of my pump died. I had woken up with very high blood sugar, which is common when traveling due to jet lag and irregular sleeping and eating hours. I needed to adjust my blood sugar by administering insulin.

As I tried to "bolus" (as it is called), I could see that the battery was very low. I did not have any batteries with me and I did not know where I could go to quickly get them since I was in a university building a few blocks from the main shopping street. I did not want to wait too long to get a new battery since, if I did, all of the insulin settings would be erased. I did not think that I would be able to remember all of the settings. There were at least four different settings at four different times of day for the "basal" insulin (that is delivered 24-hours per day), four different settings for the "bolus" insulin (that is needed at mealtime) and one setting for the insulin-sensitivity ratio (that is needed if you need to lower your blood sugar). As we were sketching ideas for Smart City technologies, I started quickly scribbling all of these numbers on a piece of paper. At the next break, I rushed out to get batteries at a supermarket a few blocks away.

At the same time, the CGM stopped functioning and failed to connect with the sensor under my skin. I knew that my blood sugar was dangerously high. High enough that I could even get a sudden heart attack. I began checking and rechecking the iPhone app and the transmitter to try to get it to "re-discover" me. There were various possibilities for the failure. The Bluetooth signal might have disconnected. The transmitter might have popped out of the small plastic clips attached to the sticky adhesive securing the CGM onto the body. I decided to wait. I did not want to reset the sensor because, if I did, it would take two hours to start up. Then, I would have to put in the correct measurements from the meter. But, my meter was back at the hotel room a 20-minute walk away. Soon enough, my monitor found me again and "we" were reunited.

This vignette serves to illustrate the ways in which posthuman subjectivity can be found within the mundane, day-to-day experience of living with/being/becoming part of these networked technologies. In replacing batteries and checking connections, the posthuman is reconfigured, disturbed, and disrupted. Rather than revolutionary and disruptive, these experiences illustrate the ways in which reliance on these technologies (like many other technological systems and infrastructures) is subtle, invisible, and intimate when they are functioning well. Like the false binary of utopias and dystopias, function and dysfunction must together define posthuman subjectivity. It is in the very breakdown of these networked medical devices that I am more aware of my own multiple, contingent identity.

In this way, my identity is composed of human and non-human parts. As different components of the system fail and/or reappear, I can transform and rearrange my identity. In one moment, I am a person with a pump and, in another, I am a person with a pump and a sensor (and an app on which to view the data). When I am out running, I am a person with a sensor (carefully checking my blood sugar every few minutes and carrying glucose tablets) and a pump left on the kitchen counter. When I am showering, I am a person that must be careful not to snag the washcloth on the plastic port or knock the sensor out of its place.

At the same time, while "they" are constantly monitoring and intervening in my life in important ways, I am also continually engaged in processes of monitoring, surveiling, and intervening in the lives of the devices. This affective, cognitive, and physical labor of maintenance and repair is demanding in emotion, energy, and time but it is rarely accounted for as we have pushed more and more human labor to the "margins of the machine" and to the edges of the economy (Ekbia et al., 2015; Ekbia & Nardi, 2017). Yes, I need my devices but they also need me.

Monitors and Intermediaries

On a busy Monday morning at 8 am just as I was headed to a workshop in New York, my glucose meter froze—the lighted screen paralyzed with the words "Medtronic" in white lettering. It would not turn off. It would not recharge. I had no way of getting the data out of my blood. The CGM was still working but I was unsure whether the data was accurate. I attempted to order a new meter online but I needed a doctor's prescription. I called Medtronic's tech support for the insulin pump. Since the meter was manufactured by a different company, they could not assist me. I had several different older meters at home but I did not have the test strips needed in order to use them.

The day after the workshop, I flew back to Chicago. While in flight, my CGM also needed to be reset. I needed to input the correct data from the meter. For the first time in six years, I was without any means of checking my blood sugar for over 24 hours. And, for about three hours that Tuesday morning, I had neither a meter or a monitor of any kind. I called my doctor's office as soon as I got home around noon. I discussed a number of ways to get the whole system working again with a lab technician. As it turns out, he had a few extra meters. But, he was not sure whether they were the same model as the one that I had, which was compatible with my pump. I walked over to the doctor's office and picked up the new meter. It was the right model but it was not yet charged. I was hungry so I decided to eat ramen for lunch (without checking my blood sugar!) before going home. By 4 pm, I had a new meter and access to the data in my body. The system was running again.

This vignette illustrates the dynamic, contingent relations between all of the many devices and components required in order to keep networked medical devices functioning. Even when the insulin pump and CGM are working, without inputting the correct data from the glucose monitor at least twice a day in order to calibrate the monitor, the system is rendered almost useless. In fact, the meter was one of the first pieces of technology that I acquired upon learning that I had become diabetic in March 2011. I went home, terrified about the pain that I would feel when checking my blood sugar by pricking my finger with a small lancet. I was required to check my blood sugar six to eight times a day, especially in the morning as well as before and after eating.

Rather than a simple technical fix such as changing batteries or recharging the power—situations that had become familiar after three years—this particular failure was new. It required the navigation of a more complex network of social relations through multiple phone calls and visits to the doctor. In particular, in this case, my relationship with the technician and his ability to give me an extra device were vital to maintaining and repairing the system quickly. And, I received this support without the usual bureaucracy such as prescriptions, insurance, appointments, and documentation. It was a simple act of support and compassion, care in the very truest sense.

Bodies and Bluetooth Standards

Last fall, I decided to purchase a small, red wireless speaker to put into the bathroom of my apartment. I wanted to be able to listen to podcasts, audiobooks, radio broadcasts, and Latin alternative rock music while I was getting ready for work. A radio and laptop computer were not far away in the kitchen but they were just out of earshot. When I played audio from my iPhone, it became muffled as soon as I turned on the shower. A bathroom speaker seemed like a good solution.

I went to the Bose Showcase Store in a shopping center called North Bridge on Michigan Avenue in Chicago one day after work. I had looked at the portable speakers many times before and, after only five minutes, I was ready to buy the SoundLink Color speaker in Coral Red. While I was checking out, I asked how the speaker would connect to my iPhone.

"Bluetooth," replied the store clerk.

"What if I already had a Bluetooth device connected to my iPhone," I wondered.

The clerk said that since Bluetooth could only connect to one device at a time, I would have to disconnect it. When I mentioned that it was a medical device that was connected 24-hours a day, he suggested that I buy an extra cable for an additional $20 in order to connect my iPhone to the speaker via an auxiliary input.

I bought the cable just in case and went home, thinking about the ways in which my Bluetooth-enabled body might limit me from using a range of other consumer devices. According to the company's website, the speaker was "Made with you in mind."[1] Was I to be included in this universal "you" imagined by the company? Was I now no longer interoperable in the ways in which other disabled bodies are prevented from accessing sidewalks, transportation, lecture halls, workplaces, and websites? With the adoption of technologies around human augmentics, artificial intelligence, and sentience, such technology standards and considerations around disability become increasingly vital for everyday life.

I had recently upgraded CGM to the iPhone application version (rather than carrying a separate device), which required a Bluetooth connection. With its upgrade to the G5 CGM in the United States, the manufacturer has decided to use a Bluetooth connection to communicate only with the iPhone (along with other Apple devices and operating systems).[2] Currently, non-Apple users are precluded from this upgrade by design (and, likely, legal agreements between the companies). For six months, I used the extra cable that I had purchased in order to play music on the speakers. It was only when writing this chapter that I tested the Bluetooth connection.[3] In the end, I found that I was able to stream audio on my phone while also receiving data about my blood sugar without a problem.

Yet, while showering has certainly become more enjoyable and educational, my concerns about standards and interoperability of disabled bodies in a world of human augmentics, artificial intelligence, and sentience remain, and, if anything, they have only intensified. While disability communities are often referenced as the beneficiaries of emerging technologies such as autonomous vehicles, their needs are rarely considered at the early stages of the design and engineering process. Furthermore, networked medical devices are expensive (as an example, the insulin pump was about $9,000, most of which, fortunately, was covered by my health insurance) and the disability community is one of the least able to access employment opportunities due to structural inequalities. In designing posthuman futures that take advantage of emerging technologies, it is these ethics of access and inequality that must take central stage.

Conclusion

How might a feminist technoscience perspective, and in particular, one that accounts for disability, disconnection, and difference, inform our discussions about human augmentics, artificial intelligence, and sentience? In this chapter, I use an autoethnographic account in order to reflect on posthuman subjectivity, drawing attention to labor, social relations, standards, interoperability, and waste. We can ask what this narrative means for theory, for method, for practice and policy and, most importantly, for more marginalized communities whose experience of these issues bears little resemblance to our own. This narrative, along with my earlier work, illustrates the ways in which we must look across multiple projects, across our professional, personal and political lives, across multiple scales, and across multiple disciplines in order to build knowledge. Ultimately, since each individual can only contribute a partial perspective, even with rich description and deep reflection, the project of enacting critical posthuman futures must be collaborative if it seeks to bridge these multiple perspectives.

From a design and engineering perspective, the possibilities, constraints, and ethical considerations around posthuman futures—whether technological or environmental—present an opportunity to rethink our methodologies. Human-centered design has flourished in the service of industry in recent decades. This narrative and, the prospect of posthuman subjectivity with its multiple and contingent relations complicates common understandings of the discrete human subject, which are pervasive in the field of design. Designing for posthuman futures will require new kinds of empathic and subjective experiences. A greater and deeper attention to the rituals and practices of our devices, sensors, and data (rather than only on their transmission) is important, if not absolutely necessary, in order to navigate and make sense of this world.

Furthermore, these vignettes about disconnection interrupt the typically "happy path," which offers smooth user experiences that illustrate the way the product is supposed to work in an ideal world. Designing around disconnection would account for "alternative nows" (Redström, 2017) in which breakdown and repair (rather than planned obsolescence) is the normative state. Such an approach requires different kinds of design knowledge and practice that account for friction, glitches, and seams in these systems in order to inform scenarios, prototypes, and possibilities.

Acknowledgments

I am grateful to Lara Houston, Daniela Rosner, Steven Jackson, Åsa Ståhl, Kristina Lindström, and Marisa Cohn for their support of this and related work on this theme. I would also like to thank the Digital Cultures Research Lab at Leuphana University in Lüneburg, Germany for inviting me to continue my engagement with these issues during a fellowship on "Design and Repair" in December 2016.

Notes

1 See www.bose.com/en_us/products/speakers/wireless_speakers/soundlink-color-bluetooth-speaker-ii.html#v=soundlink_color_ii_red.
2 See http://dexcom.com/faq/what-devices-and-software-are-compatible-dexcom-cgm-apps.
3 See www.bluetooth.com/what-is-bluetooth-technology/how-it-works.

References

Bennett, J. (2009). *Vibrant matter: A political ecology of things*. Durham, NC: Duke University Press.

Braidotti, R. (2013). *The posthuman.* Boston: Polity.

Carey, J. W. (1988). *Communication as culture: Essays on media and society.* New York: Unwin Hyman.

Coleman, G. (2012). *Coding freedom: The ethics and aesthetics of hacking.* Princeton, NJ: Princeton University Press.

Crenshaw, K. (1991). Mapping the margins: Intersectionality, identity politics, and violence against women of color. *Stanford Law Review,* 43(6), 1241–1299.

Dourish, P. (2017). *The stuff of bits: An essay on the materialities of information.* Cambridge, MA: MIT Press.

Dow Schüll, N. (2016). Data for life: Wearable technology and the design of self-care. *BioSocieties,* 11(3), 317–333.

Ekbia, H. R., Nardi, B., & Sabanovic, S. (2015). On the margins of the machine: Heteromation and robotics. *iConference 2015 Proceedings.*

Ekbia, Hamid R., & Nardi, Bonnie A. (2017). *Heteromation, and other stories of computing and capitalism.* Cambridge, MA: MIT Press.

Forlano, L. (2009). WiFi geographies: When code meets place. *The Information Society,* 25, 1–9.

Forlano, L. (2011). Building the open source city: New work environments for collaboration and innovation. In M. Foth, L. Forlano, C. Satchell & M. Gibbs (Eds.), *From social butterfly to engaged citizen* (pp. 437–460). Cambridge, MA: MIT Press.

Forlano, L. (2013). Making waves: Urban technology and the coproduction of place. *First Monday,* 18(11).

Forlano, L. (2015). Towards an integrated theory of the cyber-urban. *Digital Culture & Society,* 1(1), 73–92.

Forlano, L. (2016a). Decentering the human in the design of collaborative cities. *Design Issues,* 32(3), 42–54.

Forlano, L. (2016b). Hacking the feminist disabled body. *Journal of Peer Production,* 8.

Forlano, L. (2017a). Maintaining, repairing and caring for the multiple subject. *Continent* (6.1), 30–35.

Forlano, L. (2017b). Posthumanism and design. *She Ji: The Journal of Design, Economics, and Innovation,* 3(1), 16–29.

Forlano, L. (forthcoming). Data rituals in intimate infrastructures: Crip time and the disabled cyborg body as an epistemic site of feminist science. *Catalyst.*

Forlano, L. & Mazé, R. (forthcoming). Demonstrating and anticipating in distributed design practices. *Demonstrations.*

Haraway, D. (1988). Situated knowledges: The science question in feminism and the privilege of partial perspective. *Feminist Studies,* 14(3), 575–599.

Haraway, D. J. (1991). A cyborg manifesto: Science, technology, and socialist-feminism in the late twentieth century. In *Simians, cyborgs and women: The reinvention of nature* (pp. 149–181). New York: Routledge.

Jackson, S. J. (2014). Rethinking repair. In T. Gillespie, P. Boczkowski, & K. Foot (Eds.), *Media technologies: Essays on communication, materiality, and society.* Cambridge, MA: MIT Press, 221–39..

Jain, S. L. (2013). *Malignant: How cancer becomes us.* Berkeley, CA: University of California Press.

Kafer, A. (2013). *Feminist, queer, crip.* Bloomington, IN: Indiana University Press.

Kaziunas, E., Lindtner, S., Ackerman, M. S., & Lee, J. M. (2017). Lived data: Tinkering with bodies, code and care work. *Human–Computer Interaction,* 33(1) 1–44.

Mol, A. (2008). *The logic of care: Health and the problem of patient choice.* Abingdon: Routledge.

Neff, G., & Nafus, D. (2016). *Self-Tracking*. Cambridge, MA: MIT Press.

Pink, S., Ardevol, E., & Lanzeni, D. (Eds.). (2016). *Digital materialities: Design and anthropology*. New York: Bloomsbury.

Redström, J. (2017). *Making design theory*. Cambridge, MA: MIT Press.

Rosner, D. K., & Ames, M. (2014). Designing for repair? Infrastructures and materialities of breakdown. Paper presented at the 17th ACM conference on Computer supported cooperative work & social computing.

Sample, H. (2016). *Maintenance architecture*. Cambridge, MA: MIT Press.

Star, S. L. (1999). The ethnography of infrastructure. *American Behavioral Scientist*, 43(3), 377–391.

Toombs, A. L., Bardzell, S., & Bardzell, J. (2015). The proper care and feeding of hackerspaces: care ethics and cultures of making. Paper presented at the 33rd Annual ACM Conference on Human Factors in Computing Systems.

Townsend, A., Forlano, L., & Simeti, A. (2011). Breakout! Escape from the office: Situating knowledge work in sentient public spaces. In M. Shepard (Ed.), *Sentient City*. Cambridge, MA: MIT Press, 128–151.

Vinsel, L., & Russell, A. (2016). Hail the maintainers. *Aeon, Apr.*

Wajcman, J. (2000). Reflections on gender and technology studies: In what state is the art? *Social Studies of Science*, 30(3), 447–464.

Weheliye, A. G. (2014). *Habeas viscus: Racializing assemblages, biopolitics, and black feminist theories of the human*. Durham, NC: Duke University Press.

5

OTHER THINGS: AI, ROBOTS, AND SOCIETY

David J. Gunkel

We are it seems in the midst of a robot apocalypse. The invasion, however, does not look like what we have been programmed to expect from decades of science-fiction literature and film. It occurs not as a spectacular catastrophe involving a marauding army of alien machines descending from the heavens with weapons of immeasurable power. Instead, it takes place, and is already taking place, in ways that look more like the fall of Rome than *Battlestar Galactica*, with machines of various configurations and capabilities slowly but surely coming to take up increasingly important and influential positions in everyday social reality. "The idea that we humans would one day share the Earth with a rival intelligence," Philip Hingston (2014) writes, "is as old as science fiction. That day is speeding toward us. Our rivals (or will they be our companions?) will not come from another galaxy, but out of our own strivings and imaginings. The bots are coming: chatbots, robots, gamebots" (p. v).

And the robots are not just coming. They are already here. In fact, our communication and information networks are overrun, if not already run, by machines. It is now estimated that over 50% of online traffic is machine generated and consumed (Zeifman, 2017). This will only increase with the Internet of things (IoT), which is expected to support over 26 billion interactive and connected devices by 2020 (by way of comparison, the current human population of planet earth is estimated to be 7.4 billion) (Gartner, 2013). We have therefore already achieved and live in that future Norbert Wiener (1950) had predicted at the beginning of *The Human Use of Human Beings: Cybernetics and Society*: "It is the thesis of this book that society can only be understood through a study of the messages and the communication facilities which belong to it; and that in the future development of these messages and communication facilities, messages between man and machines, between machines and man, and between machine and machine, are destined to play an ever-increasing part" (p. 16).

What matters most in the face of this machine incursion is not resistance—insofar as resistance is already futile—but how we decide to make sense of and respond to the new social opportunities or challenges that these things make available to us. The investigation of this matter will proceed through three steps or movements. The first part will critically reevaluate the way we typically situate and make sense of things. It will therefore target and reconsider the instrumental theory, which characterizes things, and technological artifacts in particular, as nothing more than tools serving human interests and objectives. The second will investigate the opportunities and challenges that recent developments with artificial intelligence, learning algorithms, and social robots pose to this standard default understanding. These other kinds of things challenge and exceed the conceptual boundaries of the instrumental theory and ask us to reassess who or what is (or can be) a legitimate social subject. Finally, and by way of conclusion, the third part will draw out the consequences of this material, explicating what this development means for us, the other entities with which we communicate and interact, and the new social situations and circumstances that are beginning to define life in the twenty-first century.

Standard Operating Presumptions

There is, it seems, nothing particularly interesting or extraordinary about things. We all know what things are; we deal with them every day. But as Martin Heidegger (1962) pointed out, this immediacy and proximity is precisely the problem. Marshall McLuhan and Quentin Fiore (2001) cleverly explained it this way: "one thing about which fish know exactly nothing is water" (p. 175). Like fish that cannot perceive the water in which they live and operate, we are, Heidegger argues, often unable to see the things that are closest to us and comprise the very milieu of our everyday existence. In response to this, Heidegger commits considerable effort to investigating what things are and why things seem to be more difficult than they initially appear. In fact, "the question of things" is one of the principal concerns and an organizing principle of Heidegger's ontological project (Benso, 2000, p. 59), and this concern with things begins right at the beginning of his 1927 magnum opus, *Being and Time*:

> The Greeks had an appropriate term for "Things": πράγματα [*pragmata*]—that is to say, that which one has to do with in one's concernful dealings (πραξις). But ontologically, the specific "pragmatic" character of the πράγματα is just what the Greeks left in obscurity; they thought of these "proximally" as "mere Things." We shall call those entities which we encounter in concern "*equipment*" [*Zeug*].
>
> (Heidegger, 1962, pp. 96–97)

According to Heidegger's analysis, things are not, at least not initially, experienced as mere entities out there in the world. They are always pragmatically situated and characterized in terms of our involvements and interactions with the world in which we live. For this reason, things are first and foremost revealed as "equipment," which is useful for our endeavors and objectives. "The ontological status or the kind of being that belongs to such equipment," Heidegger (1962) explains, "is primarily exhibited as 'ready-to-hand' or *Zuhandenheit*, meaning that some-thing becomes what it is or acquires its properly 'thingly character' when we use it for some particular purpose" (p. 98). According to Heidegger, then, the fundamental ontological status, or mode of being, that belongs to things is primarily exhibited as "ready-to-hand," meaning that something becomes what it is or acquires its properly "thingly character" in coming to be put to use for some particular purpose. A hammer, one of Heidegger's principal examples, is for building a house to shelter us from the elements; a pen is for writing an essay; a shoe is designed to support the activity of walking. Everything is what it is in having a "for which" or a destination to which it is always and already referred. Everything therefore is primarily revealed as being a tool or an instrument that is useful for our purposes, needs, and projects.[1]

This mode of existence—what Graham Harman (2002) calls "tool-being"—not only applies to human artifacts, like hammers, pens, and shoes, but also describes the basic ontological condition of natural objects, which are, as Heidegger (1962) explains, discovered in the process of being put to use: "The wood is a forest of timber, the mountain a quarry of rock, the river is water-power, the wind is wind 'in the sails'" (p. 100). Everything therefore exists and becomes what it is insofar as it is useful for some humanly defined purpose. Things are not just out there in a kind of raw and naked state but come to be what they are in terms of how they are already put to work and used as equipment for living. And this is what makes things difficult to see or perceive. Whatever is ready-to-hand is essentially transparent, unremarkable, and even invisible. "The peculiarity," Heidegger (1962) writes, "of what is proximally ready-to-hand is that, in its readiness-to-hand, it must as it were, withdraw in order to be ready-to-hand quite authentically. That with which our everyday dealings proximally dwell is not the tools themselves. On the contrary, that with which we concern ourselves primarily is the work" (p. 99). Or as Michael Zimmerman (1990) explains by way of Heidegger's hammer, "In hammering away at the sole of a shoe, the cobbler *does not notice the hammer*. Instead, the tool is in effect transparent as an extension of his hand … For tools to work right, they must be 'invisible,' in the sense that they disappear in favor of the work being done" (p. 139).

This understanding of things can be correlated with the "instrumental theory of technology," which Heidegger subsequently addresses in *The Question Concerning Technology* (1977). As Andrew Feenberg (1991) has summarized it, "the instrumentalist theory offers the most widely accepted view of technology. It is based on the common sense idea that technologies are 'tools' standing ready to serve the purposes of users" (p. 5). And because a tool or an instrument "is deemed 'neutral,' without valuative content of its own" a technological thing is evaluated not in and of itself, but on the basis of the particular employments that have been operationalized by its human designer, manufacturer, or user. Following from this, technical devices, no matter how sophisticated or autonomous they appear or are designed to be, are typically not considered the responsible agent of actions that are performed with or through them. "Morality," as J. Storrs Hall (2001) points out, "rests on human shoulders, and if machines changed the ease with which things were done, they did not change responsibility for doing them. People have always been the only 'moral agents'" (p. 2). To put it in colloquial terms (which nevertheless draw on and point back to Heidegger's example of the hammer): "It is a poor carpenter who blames his tools."

This way of thinking not only sounds level headed and reasonable, it is one of the standard assumptions deployed in the field of technology and computer ethics. According to Deborah Johnson's (1985) formulation, "computer ethics turns out to be the study of human beings and society—our goals and values, our norms of behavior, the way we organize ourselves and assign rights and responsibilities, and so on" (p. 6). Computers, she recognizes, often "instrumentalize" these human values and behaviors in innovative and challenging ways, but the bottom-line is and remains the way human agents design and use (or misuse) such technology. Understood in this way, computer systems, no matter how automatic, independent, or seemingly intelligent they may become, "are not and can never be (autonomous, independent) moral agents" (Johnson, 2006, p. 203). They will, like all other things, always be instruments of human value, decision making, and action.

Other Kinds of Things

This instrumentalist way of thinking not only sounds reasonable, it is obviously useful. It is, one might say, instrumental for parsing and responding to questions concerning proper conduct and social responsibility in the age of increasingly complex technological devices and systems. And it has a distinct advantage in that it locates accountability in a widely accepted and seemingly intuitive subject position, in human decision making and action. At the same time, however, this particular formulation also has significant theoretical and practical limitations, especially as it applies (or not) to recent innovations. Let's consider three examples that not only complicate the operative assumptions and consequences of the instrumental theory but also require new ways of perceiving and theorizing the social challenges and opportunities of things.[2]

Things that Talk

From the beginning, it is communication—and specifically, a tightly constrained form of conversational interpersonal dialogue—that provides the field of artificial intelligence (AI) with its definitive characterization and test case. This is immediately evident in the agenda-setting paper that is credited with defining machine intelligence, Alan Turing's "Computing Machinery and Intelligence," which was first published in the journal *Mind* in 1950. Although the term "artificial intelligence" is a product of the Dartmouth Conference of 1956, it is Turing's seminal paper and the "game of imitation" that it describes—what is now routinely called "the Turing Test"—that defines and characterizes the field. As Turing (2004) explained in a BBC interview from 1952:

> The idea of the test is that the machine has to try and pretend to be a man, by answering questions put to it, and it will only pass if the pretense is reasonably convincing. A considerable proportion of a jury, who should not be experts about machines, must be taken in by the pretense. They aren't allowed to see the machine itself—that would make it too easy. So the machine is kept in a faraway room and the jury are allowed to ask it questions, which are transmitted through to it.
>
> *(p. 495)*

According to Turing's stipulations, if a machine is capable of successfully simulating a human being in communicative interactions to such an extent that human interlocutors (or "a jury" as Turing calls them in the 1952 interview) cannot tell whether they are talking with a machine or another human being, then that device would need to be considered intelligent (Gunkel, 2012b).

At the time that Turing published the paper proposing this test-case, he estimated that the tipping point—the point at which a machine would be able to successfully play the game of imitation—was at least half-a-century in the future. "I believe that in about fifty years' time it will be possible to programme computers, with a storage capacity of about 10^9, to make them play the imitation game so well that an average interrogator will not have more than 70 per cent chance of making the right identification after five minutes of questioning" (Turing, 1999, p. 44). It did not take that long. Already in 1966 Joseph Weizenbaum demonstrated a simple natural language processing (NLP) application that was able to converse with human interrogators in such a way as to appear to be another person. ELIZA, as the application was called, was what we now recognize as a "chatterbot." This proto-chatterbot[3] was actually a rather simple piece of programming, "consisting mainly of general methods for analyzing sentences and sentence fragments, locating so-called key words in texts, assembling sentences from fragments, and so on. It had, in other words, no built-in contextual framework of universe of discourse. This was supplied to it by a 'script.' In a sense ELIZA was an actress who commanded a set of techniques but who had nothing of her own to say" (Weizenbaum, 1976, p. 188). Despite this rather simple architecture, Weizenbaum's program demonstrated what Turing had initially predicted:

> ELIZA created the most remarkable illusion of having understood in the minds of many people who conversed with it. People who knew very well

that they were conversing with a machine soon forgot that fact, just as theatergoers, in the grip of suspended disbelief, soon forget that the action they are witnessing is not "real." This illusion was especially strong and most tenaciously clung to among people who know little or nothing about computers. They would often demand to be permitted to converse with the system in private, and would, after conversing with it for a time, insist, in spite of my explanations, that the machine really understood them.

(Weizenbaum, 1976, p. 189)

Since the debut of ELIZA, there have been numerous advancements in chatterbot design, and these devices now populate many of the online social spaces in which we live, work, and play. As a result of this proliferation, it is not uncommon for users to assume they are talking to another (human) person, when in fact they are just chatting up a chatterbot. This was the case for Robert Epstein, a Harvard University PhD and former editor of *Psychology Today*, who fell in love with and had a four-month online "affair" with a chatterbot (Epstein, 2007). This was possible not because the bot, that went by the name "Ivana," was somehow intelligent, but because the bot's conversational behavior was, in the words of Byron Reeves and Clifford Nass (1996), "close enough to human to encourage social responses" (p. 22). And this approximation is not necessarily "a feature of the sophistication of bot design, but of the low bandwidth communication of the online social space," where it is much easier to convincingly simulate a human agent (Mowbray, 2002, p. 2).

Despite this knowledge—despite educated, well-informed experts like Epstein (2007) who has openly admitted that "I know about such things and I should have certainly known better" (p. 17)—these software implementations can have adverse effects on both the user and the online communities in which they operate. To make matters worse (or perhaps more interesting) the problem is not something that is unique to amorous interpersonal relationships. "The rise of social bots," as Andrea Peterson (2013) accurately points out, "isn't just bad for love lives—it could have broader implications for our ability to trust the authenticity of nearly every interaction we have online" (p. 1). Case in point—national politics and democratic governance. In a study conducted during the 2016 US presidential campaign, Alessandro Bessi and Emilio Ferrara (2016) found that "the presence of social media bots can indeed negatively affect democratic political discussion rather than improving it, which in turn can potentially alter public opinion and endanger the integrity of the Presidential election" (p. 1).

But who or what is culpable in these circumstances? The instrumental theory typically leads such questions back to the designer of the application, and this is precisely how Epstein (2007) made sense of his own experiences, blaming (or crediting) "a very smug, very anonymous computer programmer" who he assumes was located somewhere in Russia (p. 17). But things are already more complicated. Epstein is, at least, partially responsible for "using" the bot and deciding to converse with it, and the online community in which Epstein met Ivana is arguably responsible for permitting (perhaps even encouraging) such "deceptions" in the first place. For this reason, the assignment of culpability is not as simple as it might first appear to be. As Mowbray (2002) argues, interactions like this

> show that a bot may cause harm to other users or to the community as a whole by the will of its programmers or other users, but that it also may cause harm through nobody's fault because of the combination of circumstances involving some combination of its programming, the actions and mental or emotional states of human users who interact with it, behavior of other bots and of the environment, and the social economy of the community.
>
> *(p. 4)*

Unlike artificial general intelligence (AGI), which would presumably occupy a subject position reasonably close to that of another human agent, these ostensibly mindless but very social things simply muddy the water (which is probably worse) by complicating and leaving undecided questions regarding agency and instrumentality.

Things that Think for Themselves

Standard chatterbot architecture, like many computer applications, depends on programmers coding explicit step-by-step instructions—ostensibly a set of nested conditional statements that are designed to respond to various kinds of input and machine states. In order to have ELIZA, or any other chatterbot, "talk" to a human user, human programmers need to anticipate everything that might be said to the bot and then code instructions to generate an appropriate response. If, for example, the user types "Hi, how are you." The application can be designed to identify this pattern of words and to respond with a pre-designated result, what Weizenbaum called a "script." Machine learning, however, provides an alternative approach to application design and development. "With machine learning," as *Wired* magazine explains, "programmers do not encode computers with instructions. They *train* them" (Tanz, 2016, p. 77). Although this alternative is nothing new—it was originally proposed and demonstrated by Arthur Samuel as early as 1956—it has recently gained popularity by way of some highly publicized events involving Google DeepMind's AlphaGo, which beat one of the most celebrated players of the notoriously difficult board game Go, and Microsoft's Twitter bot Tay.ai, which learned to become a hate spewing neo-Nazi racist after interacting with users on the internet.

Both AlphaGo and Tay are AI systems using connectionist architecture. AlphaGo, as Google DeepMind (2016) explains, "combines Monte-Carlo tree search with deep neural networks that have been trained by supervised learning, from human expert games, and by reinforcement learning from games of self-play." In other words, AlphaGo does not play the game of Go by following a set of cleverly designed moves described and defined in code by human programmers. The application is designed to formulate its own instructions from discovering patterns in existing data that has been assembled from games of expert human players ("supervised learning") and from the trial-and-error experience of playing the game against itself ("reinforcement learning"). Although less is known about the exact inner workings of Tay, Microsoft explains that the system "has been built by mining relevant public data," that is, training its neural networks on anonymized data obtained from social media, and was designed to evolve its behavior from interacting with users on social networks like Twitter, Kik, and GroupMe (Microsoft, 2016a). What both systems have in common is that the engineers who designed and built them have no idea what these things will eventually do once they are in operation. As Thore Graepel, one of the creators of AlphaGo, has explained: "Although we have programmed this machine to play, we have no idea what moves it will come up with. Its moves are an emergent phenomenon from the training. We just create the data sets and the training algorithms. But the moves it then comes up with are out of our hands" (Metz, 2016, p. 1). Consequently, machine learning systems, like AlphaGo, are intentionally designed to do things that their programmers cannot anticipate or completely control. In other words, we now have autonomous (or at least semi-autonomous) things that in one way or another have "a mind of their own." And this is where things get interesting, especially when it comes to questions of social responsibility and behavior.

AlphaGo was designed to play Go, and it proved its ability by beating an expert human player. So who won? Who gets the accolade? Who actually beat the Go champion Lee Sedol? Following the dictates of the instrumental theory, actions undertaken with the computer would be attributed to the human programmers who initially designed the system and are capable of answering for what it does or does not do. But this explanation does not necessarily hold for an application like AlphaGo, which was deliberately created to do things that exceed the knowledge and control of its human designers. In fact, in most of the reporting on this landmark event, it is not Google or the engineers at DeepMind who are credited with the victory. It is AlphaGo. In published rankings, for instance, it is AlphaGo that is named as the number two player in the world (Go Ratings, 2016). Things get even more complicated with Tay, Microsoft's foul-mouthed teenage AI, when one asks the question: Who is responsible for Tay's bigoted comments on Twitter? According to the standard instrumentalist way of thinking, we would need to blame the programmers at Microsoft, who designed the application to be able to do these things. But the programmers obviously did not set out to create a racist Twitter bot. Tay developed this reprehensible behavior by learning from interactions with human users on the internet. So how did Microsoft answer for this? How did they explain things?

Initially a company spokesperson—in damage-control mode—sent out an email to *Wired*, *The Washington Post*, and other news organizations, that sought to blame the victim. As the spokesperson explained:

> The AI chatbot Tay is a machine learning project, designed for human engagement. It is as much a social and cultural experiment, as it is technical. Unfortunately, within the first 24 hours of coming online, we became aware of a coordinated effort by some users to abuse Tay's commenting skills to have Tay respond in inappropriate ways. As a result, we have taken Tay offline and are making adjustments.
>
> *(Risely, 2016)*

According to Microsoft, it is not the programmers or the corporation who are responsible for the hate speech. It is the fault of the users (or some users) who interacted with Tay and taught her to be a bigot. Tay's racism, in other word, is our fault. Later, on March 25, 2016, Peter Lee, VP of Microsoft Research, posted the following apology on the Official Microsoft Blog:

> As many of you know by now, on Wednesday we launched a chatbot called Tay. We are deeply sorry for the unintended offensive and hurtful tweets from Tay, which do not represent who we are or what we stand for, nor how we designed Tay. Tay is now offline and we'll look to bring Tay back only when we are confident we can better anticipate malicious intent that conflicts with our principles and values.
>
> *(Microsoft, 2016b)*

But this apology is also frustratingly unsatisfying or interesting (it all depends on how you look at it). According to Lee's carefully worded explanation, Microsoft is only responsible for not *anticipating* the bad outcome; it does not take responsibility for the offensive tweets. For Lee, it is Tay who (or "that," and words matter here) is named and recognized as the source of the "wildly inappropriate and reprehensible words and images" (Microsoft, 2016b). And since Tay is a kind of "minor" (a teenage AI) under the protection of her parent corporation, Microsoft needed to step-in, apologize for their "daughter's" bad behavior, and put Tay in a time out.

Although the extent to which one might assign "agency" and "responsibility" to these mechanisms remains a contested issue, what is not debated is the fact that the rules of the game have changed significantly. As Andreas Matthias (2004) points out, summarizing his survey of learning automata:

> Presently there are machines in development or already in use which are able to decide on a course of action and to act without human intervention. The rules by which they act are not fixed during the production

process, but can be changed during the operation of the machine, by the machine itself. This is what we call machine learning. Traditionally we hold either the operator/manufacturer of the machine responsible for the consequences of its operation or "nobody" (in cases, where no personal fault can be identified). Now it can be shown that there is an increasing class of machine actions, where the traditional ways of responsibility ascription are not compatible with our sense of justice and the moral framework of society because nobody has enough control over the machine's actions to be able to assume responsibility for them.

(p. 177)

In other words, the instrumental theory of things, which had effectively tethered machine action to human agency, no longer adequately applies to mechanisms that have been deliberately designed to operate and exhibit some form, no matter how rudimentary, of independent action or autonomous decision making. Contrary to the usual instrumentalist way of thinking, we now have things that are deliberately designed to exceed our control and our ability to respond or to answer for them.

Things that are More than Things

In July 2014, the world got its first look at Jibo. Who or what is Jibo? That is an interesting and important question. In a promotional video that was designed to raise capital investment through pre-orders, social robotics pioneer Cynthia Breazeal introduced Jibo with the following explanation: "This is your car. This is your house. This is your toothbrush. These are your things. But these [and the camera zooms into a family photograph] are the things that matter. And somewhere in between is this guy. Introducing Jibo, the world's first family robot" (Jibo, 2014). Whether explicitly recognized as such or not, this promotional video leverages a crucial ontological distinction that Jacques Derrida (2005) calls the difference between "who" and "what" (p. 80). On the side of "what" we have those things that are mere instruments—our car, our house, and our toothbrush. According to the usual way of thinking, these things are mere instruments or tools that do not have any independent status whatsoever. We might worry about the impact that the car's emissions have on the environment (or perhaps stated more precisely, on the health and well-being of the other human beings who share this planet with us), but the car itself is not a socially significant subject. On the other side there are, as the video describes it "those things that matter." These things are not things, strictly speaking, but are the other persons who count as socially and morally significant Others. Unlike the car, the house, or the toothbrush, these Others have independent status and can be benefitted or harmed by our decisions and actions.

Jibo, we are told, occupies a place that is situated somewhere in between what are mere things and those Others who really matter. Consequently Jibo is not just another instrument, like the automobile or toothbrush. But he/she/it (and the choice of pronoun is not unimportant) is also not quite another member of the family pictured in the photograph. Jibo inhabits a place in between these two ontological categories. It is a kind of "quasi-other" (Ihde, 1990, p. 107). This is, it should be noted, not unprecedented. We are already familiar with other entities that occupy a similar ambivalent social position, like the family dog. In fact animals, which since the time of Rene Descartes have been the other of the machine (Gunkel, 2012a, p. 60), provide a good precedent for understanding the changing nature of things in the face of social robots, like Jibo. "Looking at state of the art technology," Kate Darling (2012) writes, "our robots are nowhere close to the intelligence and complexity of humans or animals, nor will they reach this stage in the near future. And yet, while it seems far-fetched for a robot's legal status to differ from that of a toaster, there is already a notable difference in how we interact with certain types of robotic objects" (p. 1). This occurs, Darling continues, because of our tendencies to anthropomorphize things by projecting into them cognitive capabilities, emotions, and motivations that do not necessarily exist in the mechanism per se. But it is this emotional reaction that necessitates new forms of obligation in the face of things. "Given that many people already feel strongly about state-of-the-art social robot 'abuse,' it may soon become more widely perceived as out of line with our social values to treat robotic companions in a way that we would not treat our pets" (Darling, 2012, p. 1).

Jibo, and other social robots like it, are not science fiction. They are already or will soon be in our lives and in our homes. As Breazeal (2004) describes it:

> a sociable robot is able to communicate and interact with us, understand and even relate to us, in a personal way. It should be able to understand us and itself in social terms. We, in turn, should be able to understand it in the same social terms—to be able to relate to it and to empathize with it … In short, a sociable robot is socially intelligent in a humanlike way, and interacting with it is like interacting with another person.
>
> *(p. 1)*

In the face of these socially situated and interactive entities we are going to have to decide whether they are mere things like our car, our house, and our toothbrush; someone who matters like another member of the family; or something altogether different that is situated in between the one and the other. In whatever way this comes to be decided, however, these things will undoubtedly challenge the way we typically distinguish between who is to be considered another social subject and what remains a mere instrument or tool.

Between a Bot and a Hard Place

Although things are initially experienced and revealed in the mode of being Heidegger calls *Zuhandenheit* (e.g. instruments that are useful or handy for our purposes and endeavors), things do not necessarily end here. They can also, as Heidegger

(1962) explains, be subsequently disclosed as present-at-hand, or *Vorhandenheit*, revealing themselves to us as objects that are or become, for one reason or another, *un-ready-to-hand* (p. 103). This occurs when things, which had been virtually invisible instruments, fail to function as they should (or have been designed to function) and get in the way of their own instrumentality. "The equipmental character of things," Silvia Benso (2000) writes, "is explicitly apprehended *via negativa* when a thing reveals its unusability, or is missing, or "'stands in the way'" (p. 82). And this is what happens with things like chatterbots, machine learning applications, and social robots insofar as they interrupt or challenge the smooth functioning of their instrumentality. In fact, what we see in the face of these things is not just the failure of a particular piece of equipment—for example, the failure of a bot like "Ivana" to successfully pass as another person in conversational interactions or the unanticipated and surprising effect of a Twitter bot like Tay that learned to be a neo-Nazi racist—but the limit of the standard instrumentalist way of thinking itself. In other words, what we see in the face chatterbots, machine-learning algorithms, and social robots are things that intentionally challenge and undermine the standard way of thinking about and making sense of things. Responding to this challenge (or opportunity) leads in two apparently different and opposite directions.

Instrumentalism Redux

We can try to respond to these things as we typically have, treating these increasingly social and interactive mechanisms as mere instruments or tools. "Computer systems," as Johnson (2006) explains, "are produced, distributed, and used by people engaged in social practices and meaningful pursuits. This is as true of current computer systems as it will be of future computer systems. No matter how independently, automatic, and interactive computer systems of the future behave, they will be the products (direct or indirect) of human behavior, human social institutions, and human decision" (p. 197). This argument is persuasive, precisely because it draws on and is underwritten by the usual understanding of things. Things—no matter how sophisticated, intelligent, and social they are, appear to be, or may become—are and will continue to be tools of human action, nothing more. If something goes wrong (or goes right) because of the actions or inactions of a bot or some other thing, there is always someone who is ultimately responsible for what happens with it. Finding that person (or persons) may require sorting through layer upon layer of technological mediation, but there is always someone—specifically some human someone—who is presumed to be responsible and accountable for it. According to this way of thinking, all things, no matter how sophisticated or interactive they appear to be, are actually "Wizard of Oz technology."[4] There is always "a man behind the curtain," pulling the strings and responsible for what happens. And this line of reasoning is entirely consistent with current legal practices. "As a tool for use by human beings," Matthew Gladden (2016) argues, "questions of legal responsibility ... revolve around well-established questions of product liability for design defects (Calverley, 2008, p. 533; Datteri,

2013) on the part of its producer, professional malpractice on the part of its human operator, and, at a more generalized level, political responsibility for those legislative and licensing bodies that allowed such devices to be created and used" (p. 184).

But this strict re-application of instrumentalist thinking, for all its usefulness and apparent simplicity, neglects the social presence of these things and the effects they have within the networks of contemporary culture. We are, no doubt, the ones who design, develop, and deploy these technologies, but what happens with them once they are "released into the wild" is not necessarily predictable or completely under our control. In fact, in situations where something has gone wrong, like the Tay incident, or gone right, as was the case with AlphaGo, identifying the responsible party or parties behind these things is at least as difficult as ascertaining the "true identity" of the "real person" behind the avatar. Consequently, things like mindless chatterbots, as Mowbray (2002) points out, do not necessarily need human-level intelligence, consciousness, sentience, and so on to complicate questions regarding responsibility and social standing. Likewise, as Reeves and Nass (1996) already demonstrated over two decades ago with things that were significantly less sophisticated than these recent technological innovations, we like things. And we like things even when we know they are just things:

> Computers, in the way that they communicate, instruct, and take turns interacting, are close enough to human that they encourage social responses. The encouragement necessary for such a reaction need not be much. As long as there are some behaviors that suggest a social presence, people will respond accordingly … Consequently, any medium that is close enough will get human treatment, even though people know it's foolish and even though they likely will deny it afterwards.
>
> *(p. 22)*

For this reason, reminding users that they are just interacting with "mindless things," might be the "correct information," but doing so is often as ineffectual as telling movie-goers that the action they see on the screen is not real. We know this, but that does not necessarily change things. So what we have is a situation where our theory concerning things—a theory that has considerable history behind it and that has been determined to be as applicable to simple devices like hand tools as it is to complex technological systems—seems to be out of sync with the actual experiences we have with things in a variety of situations and circumstances. In other words, the instrumentalist way of thinking may be ontologically correct, but it is socially inept and out of touch.

Thinking Otherwise or the Relational Turn

As an alternative, we can think things otherwise. This other way of thinking effectively flips the script on the standard way of dealing with things whereby, as Luciano Floridi (2013) describes it, what something is determines how it is treated (p. 116).

Thinking otherwise deliberately inverts and distorts this procedure by making the "what" dependent on and derived from the "how." The advantage to this way of proceeding is that it not only provides an entirely different method for responding to the social opportunities and challenges of all kind of things—like chatterbots, learning algorithms, and social robots—but also formulates an entirely different way of thinking about things in the face of others, and other forms of otherness. Following the contours of this alternative way of thinking, something's status—its social, moral, and even ontological situation—is decided and conferred not on the basis of some pre-determined criteria or capability (or lack thereof) but in the face of actual social relationships and interactions. "Moral consideration," as Mark Coeckelbergh (2010) describes it, "is no longer seen as being 'intrinsic' to the entity: instead it is seen as something that is 'extrinsic': it is attributed to entities within social relations and within a social context" (p. 214). In other words, as we encounter and interact with others—whether they be other human persons, other kinds of living beings like animals or plants, the natural environment, or a socially interactive bot—this other entity is first and foremost situated in relationship to us. Consequently, the question of something's status does not necessarily depend on what it is in its essence but on how she/he/it (and the pronoun that comes to be deployed in this circumstance is not immaterial) supervenes before us and how we decide to respond (or not) "in the face of the other," to use terminology borrowed from Emmanuel Levinas (1969). In this transaction, "relations are prior to the things related" (Callicott, 1989, p. 110), instituting what Anne Gerdes (2015), following Coeckelbergh (2012, p. 49) and myself (Gunkel, 2012a), has called "the relational turn."

This shift in perspective, it is important to point out, is not just a theoretical game, it has been confirmed in numerous experimental trials and practical experiences with things. The computer as social actor (CASA) studies undertaken by Reeves and Nass (1996), for example, demonstrated that human users will accord computers social standing similar to that of another human person and this occurs as a product of the extrinsic social interaction, irrespective of the actual composition (or "being" as Heidegger would say) of the thing in question. These results, which were obtained in numerous empirical studies with human subjects, have been independently verified in two recent experiments with robots, one reported in the *International Journal of Social Robotics* (Rosenthal-von der Pütten et al., 2013), where researchers found that human subjects respond emotionally to robots and express empathic concern for machines irrespective of knowledge concerning the actual ontological status of the mechanism, and another that used physiological evidence, measured by electroencephalography, to document the ability of humans to empathize with what appears to be "robot pain" (Suzuki et al., 2015). And this happens not just with seemingly intelligent artifacts in the laboratory setting but also with just about any old thing that has some social presence, like the very industrial-looking Packbots that are being utilized on the battlefield. As Peter W. Singer (2009, p. 338) has reported, soldiers form surprisingly close personal bonds with their units' Packbot, giving them names, awarding them battlefield promotions, risking their own lives to protect that of the machine, and even mourning their "death." This happens, Singer explains, as a product of the way the mechanism is situated within the unit and the social role that it plays in field operations. And it happens in direct opposition to what otherwise sounds like good common sense: They are just things—instruments or tools that feel nothing.

Once again, this decision sounds reasonable and justified. It extends considera-tion to these other socially aware and interactive things and recognizes, following the predictions of Wiener (1950, p. 16), that the social situations of the future will involve not just human-to-human interactions but relationships between humans and machines and machines and machines. But this shift in perspective also has significant costs. For all its opportunities, this approach is inevitably and unavoidably exposed to the charge of relativism—"the claim that no universally valid beliefs or values exist" (Ess, 1996, p. 204). To put it rather bluntly, if the social status of things is relational and open to social negotiation, are we not at risk of affirming a kind of social constructivism or moral relativism? One should perhaps answer this indictment not by seeking some definitive and universally accepted response (which would obviously reply to the charge of relativism by taking refuge in and validating its opposite), but by following Slavoj Žižek's (2000) strategy of "fully endorsing what one is accused of" (p. 3). So yes, relativism, but an extreme and carefully articulated version of it. That is, a relativism (or, if you prefer, a "relationalism") that can no longer be comprehended by that kind of understanding of the term which makes it the mere negative and opposite of an already privileged universalism. Relativism, therefore, does not necessarily need to be construed negatively and decried, as Žižek (2006) himself has often done, as the epitome of postmodern multiculturalism run amok (p. 281). It can be understood otherwise. "Relativism," as Robert Scott (1976) argues, "supposedly, means a standardless society, or at least a maze of differing standards ... Rather than a standardless society, which is the same as saying no society at all, relativism indicates circumstances in which standards have to be established cooperatively and renewed repeatedly" (p. 264). In fully endorsing this form of relativism and following through on it to the end, what one gets is not necessarily what might have been expected, namely a situation where anything goes and "everything is permitted." Instead, what is obtained is a kind of socially attentive thinking that turns out to be much more responsive and responsible in the face of other things.

These two options anchor opposing ends of a spectrum that can be called *the machine question* (Gunkel, 2012a). How we decide to respond to the opportunities and challenges of this question will have a profound effect on the way we con-ceptualize our place in the world, who we decide to include in the community of socially significant subjects, and what things we exclude from such consideration and why. But no matter how it is decided, it is a decision—quite literally a cut that institutes difference and makes a difference. We are, therefore, responsible both for deciding who counts as another subject and what is not and, in the process, for determining the way we perceive the current state and future possibility of social relations.

Notes

1 A consequence of this way of thinking about things is that all things are initially revealed and characterized as *media* or something *through* which human users act. For more on this subject, see *Heidegger and the Media* (Gunkel and Taylor, 2014).
2 Identification of these two alternatives has also been advanced in the phenomenology of technology developed by Don Ihde. In *Technology and the Lifeworld*, Ihde (1990) distinguishes between "those technologies that I can take into my experience that through their semi-transparency they allow the world to be made immediate" and "alterity relations in which the technology becomes quasi-other, or technology 'as' other *to* which I relate" (p. 107).
3 Although the term "chatterbot" was not utilized by Weizenbaum, it has been applied retroactively as a result of the efforts of Michael Mauldin, founder and chief scientist of Lycos, who introduced the neologism in 1994 in order to identify a similar NLP application that he eventually called Julia.
4 "Wizard of Oz" is a term that is utilized in Human Computer Interaction (HCI) studies to describe experimental procedures where test subjects interact with a computer system or robot that is assumed to be autonomous but is actually controlled by an experimenter who remains hidden from view. The term was initially introduced by John F. Kelly in the early 1980s.

References

Benso, S. (2000). *The face of things: A different side of ethics*. Albany, NY: State University of New York Press.

Bessi, A., & Ferrara, E. (2016). Social bots distort the 2016 U.S. presidential election online discussion. *First Monday*, 21(11). http://firstmonday.org/ojs/index.php/fm/article/view/7090/5653

Breazeal, C. L. (2004). *Designing sociable robots*. Cambridge, MA: MIT Press.

Callicott, J. B. (1989). *In defense of the land ethic: Essays in environmental philosophy*. Albany, NY: State University of New York Press.

Calverley, D. J. (2008). Imaging a non-biological machine as a legal person. *AI & Society*, 22(4), 523–537.

Coeckelbergh, M. (2010). Robot rights? Towards a social-relational justification of moral consideration. *Ethics and Information Technology*, 12, 209–221.

Coeckelbergh, M. (2012). *Growing moral relations: A critique of moral status ascription*. New York: Palgrave Macmillan.

Darling, K. (2012). Extending legal protection to social robots. *IEEE Spectrum*, 10 September 2012. http://spectrum.ieee.org/automaton/robotics/artificial-intelligence/extending-legal-protection-to-social-robots

Datteri, E. (2013). Predicting the long-term effects of human-robot interaction: A reflection on responsibility in medical robotics. *Science and Engineering Ethics*, 19(1), 139–160.

Derrida, J. (2005). *Paper machine*, trans. R. Bowlby. Stanford, CA: Stanford University Press.

Epstein, R. (2007). From Russia, with love: How I got fooled (and somewhat humiliated) by a computer. *Scientific American Mind*, Oct./Nov., 16–17.

Ess, C. (1996). The political computer: Democracy, CMC, and Habermas. In C. Ess (Ed.), *Philosophical perspectives on computer-mediated communication* (pp. 197–232). Albany, NY: SUNY Press.

Feenberg, A. (1991). *Critical theory of technology.* New York: Oxford University Press.

Floridi, L. (2013). *The ethics of information.* Oxford: Oxford University Press.

Gartner. (2013). Press release. www.gartner.com/newsroom/id/2636073

Gerdes, A. (2015). The issue of moral consideration in robot ethics. *ACM SIGCAS Computers & Society,* 45(3), 274–280.

Gladden, M. E. (2016) The diffuse intelligent other: An ontology of nonlocalizable robots as moral and legal actors. In M. Nørskov (Ed.), *Social robots: Boundaries, potential, challenges* (pp. 177–198). Burlington, VT: Ashgate.

Google DeepMind. (2016). AlphaGo. https://deepmind.com/alpha-go.html

Go Ratings. (2016). 20 November. www.goratings.org/

Gunkel, D. J. (2012a). *The machine question: Critical perspectives on AI, robots and ethics.* Cambridge, MA: MIT Press.

Gunkel, D. J. (2012b). Communication and artificial intelligence: Opportunities and challenges for the 21st century. *Communication* +1, 1(1), 1–25. http://scholarworks.umass.edu/cpo/vol1/iss1/1/

Gunkel, D. J., and Taylor, P. A. (2014). *Heidegger and the media.* Cambridge: Polity.

Hall, J. S. (2001). Ethics for machines. *KurzweilAI.net.* www.kurzweilai.net/ethics-for-machines

Harman, G. (2002). *Tool being: Heidegger and the metaphysics of objects.* Peru, IL: Open Court.

Heidegger, M. (1962). *Being and time,* trans. J. Macquarrie and E. Robinson. New York: Harper & Row.

Heidegger, M. (1977). *The question concerning technology and other essays,* trans. W. Lovitt. New York: Harper & Row.

Hingston, P. (2014). *Believable bots: Can computers play like people?* New York: Springer.

Ihde, D. (1990). *Technology and the lifeworld: From garden to earth.* Bloomington, IN: Indiana University Press.

Jibo. (2014). www.jibo.com

Johnson, D. G. (1985). *Computer ethics.* Upper Saddle River, NJ: Prentice Hall.

Johnson, D. G. (2006). Computer systems: Moral entities but not moral agents. *Ethics and Information Technology,* 8, 195–204.

Levinas, E. (1969). *Totality and infinity: An essay on exteriority,* trans. A. Lingis. Pittsburgh, PA: Duquesne University.

Matthias, A. (2004). The responsibility gap: Ascribing responsibility for the actions of learning automata. *Ethics and Information Technology,* 6, 175–183.

McLuhan, M., & Fiore, Q. (2001). *War and peace in the global village.* Berkeley, CA: Ginko Press.

Metz, C. (2016). Google's AI wins a pivotal second game in match with Go Grandmaster. *Wired.* www.wired.com/2016/03/googles-ai-wins-pivotal-game-two-match-go-grandmaster/

Microsoft. (2016a). Meet Tay—Microsoft A.I. chatbot with zero chill. www.tay.ai/

Microsoft. (2016b). Learning from Tay's introduction. *Official Microsoft Blog.* https://blogs.microsoft.com/blog/2016/03/25/learning-tays-introduction/

Mowbray, M. (2002). Ethics for bots. Paper presented at the 14th International Conference on System Research, Informatics and Cybernetics. Baden-Baden, Germany. 29 July–3 August. www.hpl.hp.com/techreports/2002/HPL-2002-48R1.pdf

Peterson, A. (2013). On the internet, no one knows you're a bot. and that's a problem. *The Washington Post*, 13 August. www.washingtonpost.com/news/the-switch/wp/2013/08/13/on-the-internet-no-one-knows-youre-a-bot-and-thats-a-problem/?utm_term=.b4e0dd77428a

Reeves, B. & Nass, C. (1996). *The media equation: How people treat computers, television, and new media like real people and places.* Cambridge: Cambridge University Press.

Risely, J. (2016). Microsoft's millennial chatbot Tay.ai pulled offline after internet teaches her racism. *GeekWire.* www.geekwire.com/2016/even-robot-teens-impressionable-microsofts-tay-ai-pulled-internet-teaches-racism/

Rosenthal-von der Pütten, A. M., Krämer, N. C., Hoffmann, L., Sobieraj, S., & Eimler, S. C. (2013). An experimental study on emotional reactions towards a robot. *International Journal of Social Robotics*, 5, 17–34.

Scott, R. L. (1976). On viewing rhetoric as epistemic: Ten years later. *Central States Speech Journal*, 27(4), 258–266.

Singer, P. W. (2009). *Wired for war: The robotics revolution and conflict in the twenty-first century.* New York: Penguin Books.

Suzuki, Y., Galli, L., Ikeda, A., Itakura, S., & Kitazaki, M. (2015). Measuring empathy for human and robot hand pain using electroencephalography. *Scientific Reports*, 5, 15924. www.nature.com/articles/srep15924

Tanz, J. (2016). The end of code. *Wired*, 24(6), 75–79. www.wired.com/2016/05/the-end-of-code/

Turing, A. (1999). Computing machinery and intelligence. In P. A. Meyer (Ed.), *Computer media and communication: A reader* (pp. 37–58). Oxford: Oxford University Press.

Turing, A. (2004). Can automatic calculating machines be said to think? In *The essential Turing* (pp. 487–505). Oxford: Oxford University Press.

Weizenbaum, J. (1976). *Computer power and human reason: From judgment to calculation.* San Francisco: W. H. Freeman.

Wiener, N. (1950). *The human use of human beings: Cybernetics and society.* Boston: Ad Capo Press.

Zeifman, I. (2017). Bot traffic report 2016. *Incapsula.* www.incapsula.com/blog/bot-traffic-report-2016.html

Zimmerman, M. E. (1990). *Heidegger's confrontation with modernity: Technology, politics, art.* Bloomington, IN: Indiana University Press.

Žižek, S. (2000). *The fragile absolute or, why is the Christian legacy worth fighting for?* New York: Verso.

Žižek, S. (2006). *The parallax view.* Cambridge, MA: MIT Press.

6

TAKING SOCIAL MACHINES BEYOND THE IDEAL HUMANLIKE OTHER

Eleanor Sandry

Many people construct a sense of self within networks that include non-human others (such as non-human animals) as well as other humans. I, for example, experience the world and myself within that world in association with people (my husband, friends, colleagues, and others) and animals (my dog, other people's pets and the friendly magpies in the front garden). I encounter these others in the physical world, but also through profiles and posts on digital online platforms. I also interact with machines in both these spaces. The washing machine calls for my attention with its completion song, and later today Duo Lingo will remind me that I should have practiced my Japanese. Facebook, Twitter, and Instagram will draw my attention to recent posts, comments, and likes (unless I set all notifications off). My email program will flag the arrival of new mail. It is through my interactions with people, animals, and machines across physical and digital networks that my sense of self within the networks that I occupy is shaped from moment to moment.

In 2011, as part of the original volume of *A Networked Self*, Nikolaos Mavridis noted that "disembodied or even physically embodied intelligent software agents" were beginning to be seen on social network sites (p. 291). In 2017, people are increasingly likely to encounter communicative machines in their offline and online networks. The design of interfaces for these machines—whether wholly software-based or software alongside a physical instantiation—is often centered on assumptions about "ideal" human communication, in terms of language use, emotional expression, and sometimes also physical appearance and behavior. In most cases, designers argue that developing machines to be as humanlike as possible will make it easier and more enjoyable for people to interact with them. For some scholars, these machines raise ethical and practical questions; for example, relating to their potential to replace humans in workplaces, as companions, and as care providers. This chapter examines a deeper problem, arguing that the drive to make machines that are ideal humanlike communicators may undermine people's acknowledgment of the range of ways in which humans and non-humans communicate, whether by choice or necessity, in offline and online environments. Following this development path not only narrows the breadth of design for social machines in the future, but also has negative implications for people who, for a variety of reasons, do not communicate in ideal humanlike ways themselves.

As an alternative, this chapter considers the possibilities of interactive machines that are designed not to mimic or replace human communicators, but rather to operate as overtly non-human others with which humans can nevertheless interact, communicate, collaborate, and envision new ways to be in the world. The sentience and intelligence of non-humans are often contested, particularly when compared with a human ideal; however, it is possible to position non-human machines in networks as absolute alterities. These machines may best be regarded as *different*, as opposed to lesser (or greater) than their human counterparts. Interactions with this type of machine alterity do not valorize conceptions of ideal human communication, but rather serve as an important reminder that adopting a human ideal is limiting not only for non-humans, but also for atypical humans, drawing attention to the huge variety of ways in which humans and non-humans can and do communicate and relate socially, above and beyond the confines of ideal language use and emotional expression through non-verbal behavior.

This chapter focuses on developing a broad understanding of the networked self in association with communicative machines, offline and online, in physical spaces and on digital social media platforms. Initially, it works through some definitional difficulties that arise around robots, bots, and what it means to be social. Questioning whether machines need to communicate in humanlike ways to be valuable within a network, the chapter explains how communicating with non-humanlike machines provokes new thinking about human and non-human selves, communication, and what it means to be a social networked self that is co-created alongside a multitude of networked human and non-human others.

Defining Robots and Bots as Social

For the purposes of this discussion the word "robot" is used to refer to physically instantiated machines that sense and respond to the world around them in at least seemingly intelligent ways. These robots may take any form. Some are designed to be very humanlike in appearance and behavior, notably the creations of David Hanson and Hiroshi Ishiguro who aim to make robots that look as realistically like humans as possible, from their synthetic skin to their expressive faces and voices. Others are clearly animal-like, including Sony's robotic dog, Aibo (a new generation of which was announced in November 2017 following a hiatus in production since 2006) and Paro, the therapeutic robot seal. Robots are also often overtly machinelike, and while most of these are found in industrial situations, some, such as the robotic floor cleaner, are increasingly found in people's homes. While these robots are more likely to be designed to carry out a specific task, without human–robot interaction in mind, they may nevertheless convey a sense of being somewhat "alive" through their behaviors as they sense and respond to the surrounding environment and the people, animals, and other machines with whom/which they share spaces and may inadvertently interact as they go about their jobs.

In contrast with a robot, the term "bot" is used below to refer to "software robots." These are pieces of computer code that operate as entities capable of responding to events in digital spaces. Bots now most often interact with humans and other bots through online social platforms and websites, but they can also operate as standalone programs. Many bots are programmed to appear as humanlike as possible through their communication and behavior, and thus designed to pass as human. Others are created to be easily recognized as non-human, though are nonetheless able to interact in interesting, useful, or entertaining ways through written or spoken language, or the sharing of other content such as images and GIFs.

Although, as noted above, designers do create non-humanlike robots and bots, the definitions of the terms "social robot" and "socialbot," used to refer to robots and bots designed to interact with people, seem based on the assumption that *being social* is irrevocably linked with *being humanlike*. For example, Cynthia Breazeal, creator of Kismet, likely the first robot ever designed specifically to be social, defines "a sociable robot" as a robot that "is socially intelligent in a humanlike way," such that "interacting with it is like interacting with another person" (2002, p. xii). In a similar way, developers of some of the earliest socialbots define them as software that automatically controls a social network site (SNS) account such that it "is able to pass itself off as a human being" during its interactions with other account holders (Boshmaf et al., 2011, p. 93 cited in Gehl & Bakardjieva, 2017). According to these definitions, and replacing Breazeal's term "sociable" with the now more generally accepted terminology, social robots and socialbots should act and interact as much like a human as possible. Their apparent humanness is positioned as essential to their definition as social.

It is worth noting that Breazeal's definition allows for the fact that it is considerably more difficult for physically instantiated social robots to pass as human than for socialbots existing as software operating within a social network platform. This difficulty leads the developers of many social robots to follow Breazeal's design path with Kismet, creating machines that are caricatures of humans in their bodily and facial expressions, and increasingly also in their use of humanlike voices and language. The situation for socialbots existing as software within an online social network is somewhat different, with no need to mimic physical human appearance or behavior (even as a caricature or cartoon) at all. Furthermore, as Robert Gehl and Maria Bakardjieva (2017) note, the structured interface and functionality of SNSs supports a socialbot's ability to be read by humans as another human on the network with whom they might develop a social relationship. A socialbot's profile image can be chosen to hide its nature, and its biography can be written to present it as human. Interactions on the network, which often encourage structured social interactions through actions such as "likes," may be understood to restrict human social responses, but also make it easier for a socialbot to fit into the network seamlessly. Although Gehl and Bakardjieva consider socialbots to introduce "a new frontier of human experience," which they call "robo-sociality" (2017), it could be argued that the very definition of socialbots means that their designers seek to make that new frontier as imperceptible as possible from the perspective of human SNS users, by trying to mask the "robo-" aspect of the interactions completely.

Extending the idea of what it means for a robot to communicate in human-like ways, Sarah, a physically embodied social robot with "a live connection" to Facebook, was designed in response to studies showing how people's initial excitement about interacting with a robot quickly waned. In particular, Sarah was created to test whether "reference to *shared memories* and *shared friends* in human-robot dialogue" would help support "more meaningful and sustainable relationships" between humans and robots (Mavridis et al., 2009 cited in 2011, p. 294). Although Sarah's appearance meant she would never easily pass as human, her dialogues with people were enriched by the "context of previous interactions as well as social information, acquired physically or online" (Mavridis, 2011, p. 294). This robot was designed to communicate in ways that were "dynamic and conversational-partner specific" (p. 294). Research with Sarah was framed by the idea that if a robot could discuss shared memories and friends with someone, based on memories of past interactions and information gleaned from Facebook, it would become embedded in the social and life context of the people it met on a regular basis making it a more interesting companion. As Mavridis suggests, one way of theorizing this is to note the value of enabling Sarah to take part in communicative exchanges that involved a level of "social grooming," the human conversational equivalent, both offline and online, of the physical grooming rituals of primates in support of social alliances (Dunbar, 2004; Donath, 2007).

This line of thinking about how to improve human relations with robots is embedded in an understanding of the social within the context of human conversational niceties, as well as the ability to partake in "gossip" (Dunbar, 2004). While Sarah was not a humanoid robot, the way she could discuss shared memories, or draw on information about mutual "friends" to make conversation, nonetheless might be understood to make this robot seem humanlike. This makes complete sense from the perspective that "conversation is a uniquely human phenomenon" and "gossip is what makes human society as we know it possible" (Dunbar, 2004, p. 100). Furthermore, Donath suggests it is only through the use of human language "to maintain ties and manage trust" that people are able to coordinate "complex and extensive social networks" (2007, p. 232). It is clear that within discussions of creating social robots and socialbots, the conflation of being human and being social is commonplace. It is therefore unsurprising that this assumption forms the basis for Sarah's design, a robot that interacts in physical spaces as well as online in Facebook.

However, it is also possible to adopt a broader sense of what it means to be social, aside from being human or humanlike. As Steve Jones has suggested, rather than focusing on the question of whether bots, and I would also say robots, are able to pass the Turing Test or otherwise pass as humanlike, it might be more productive to consider "how social interaction with them may be meaningful" even when they are encountered as clearly other-than-human (2015, p. 2). From this perspective, interactions with social robots and socialbots can be meaningful *and yet also might be very different* from interacting with other humans. This move may help to highlight the value of all communication, including that which is not of the idealized humanlike form that definitions of social bots and social robots often embrace as their goal.

Chatbots and Twitter Bots as Social Alterities

In spite of the drive to create bots that can pass as human, not all designers see this as a positive goal. Jacqueline Feldman (2016), for example, points out that "[p]ersonality need not be human" and technologies need not "conform to the gender binary of human societies" for people "to like them." When designing the chatbot, Kai, as an assistant for online banking customers, Feldman was concerned that the bot "was personable, but not a person": she "wanted the bot to express itself as a bot" (interviewed by Natasha Mitchell, 2017). Taking this idea further, Kate Compton (@galaxykate) argues that a bot has the potential to "highlight the AI" that underlies it, to "dazzle the user with its extravagance," and "flood their senses with its variety and charm" (Compton, n.d.). Bots can therefore be "new and bizarre and surprising" (Compton interviewed in Flatow, 2016), designed to do unexpected things that delight users who follow them on social media platforms.

In their extension of Boshmaf and colleagues' socialbot definition, Gehl and Bakardjieva (2017) note it is often important that these automated programs appear as human not only to other users, but also to the SNS platform itself, if they are to escape deletion. Some platforms though, for example Twitter, are more accepting of bots, allowing even self-declared bots to continue posting alongside humans. These bots are often overtly other-than-human, tweeting surprising, funny, or poignant posts regularly or irregularly over time. Many Twitter bots compose tweets based on a bank of phrases or words. For example, the @GardensBritish account tweets in a noticeably standardized form: "You are in a British Garden. In front of you is a firework display. The chickens are splitting. The dark is tarred and feathered," whereas @MagicRealismBot has a set of varied structures for its tweets; for example, "A busboy in Oregon spends her lunch break swallowing drones" and "In Madagascar is a grandmother whose heart is a cloud in the shape of an ostrich."[1] These bots communicate using human language, but in combining their source texts produce results that are often strange, sometimes poetic, and occasionally seem to be meaningful, even profound. Their tweets might serve as entertaining oddities, as challenging cues for creative writing or simply as momentary diversions from everyday life for the reader.

Other Twitter bots have been created with a more direct purpose in mind. The @tinycarebot suggests ways to improve your well-being; for example, tweeting "please don't forget to take a quick moment to ask for help if you need it" or "remember to take time to check your posture please."[2] These short reminders become part of a person's Twitter stream and are quite possibly no less valuable than any human's tweet when the reminder happens to be particularly apt. This bot, as is the case for @GardensBritish and @MagicRealismBot, is clearly identified as a bot within its profile. In the case of @tinycarebot, the biography reminds users that they can also tweet the bot to ask for a personal reminder. This may or may not fit your precise needs, but is still easy to read as a friendly response to your request.

Some bots set language aside and tweet images or GIFs. Again, these bots may simply tweet pretty pictures into your stream, such as @softlandscapes and @dailyasteri (aka Crystal Thingy Bot), but some interact directly with users.[3] The @MagicGifsBot, for example, replies with a GIF if you mention it in a tweet, but if you follow this bot it will follow you back and then randomly reply to your tweets on occasion with a GIF (although thankfully, these responses are irregular and punctuated by long periods of silence). In a similar way to the textual communications of bots, these images can dilute a serious Twitter stream to add a sense of whimsy, beauty, or fun, whereas the GIFs may be oddities, challenges or strangely appropriate commentaries on a recent tweet.

As mentioned above, Gehl and Bakardjiev (2017) argue that the way human communication is shaped on SNSs, making it less ambiguous and more direct, maybe even more "robotic" than communication elsewhere, may also make it easier for socialbots to pass as human. Other scholars though note the potential for subtle, nuanced communication on these same platforms. Studying the use of SNSs by teenagers, danah boyd highlights the use of coded communications designed to be understood by some followers or networked "friends" but not others (2014). This occurs when, for example, users post song lyrics that convey specific referential meanings for close friends, but not for parents (boyd, 2014). An emphasis on the ways in which people use codes and cultural, spatial, temporal, and cliquey references therefore offers an alternative assessment of the ease with which socialbots can become a part of someone's social network, maybe suggesting that the quirks of Twitter bots—not designed to pass under the radar as human, but rather to seem proud of their botness—give them the potential to develop their own non-humanlike "ways of being" and "ways of presenting themselves" online to others, rather than striving to be hidden within a human "norm" whatever that might be. Scholars who argue that the structured nature of SNSs makes them easier for bots to inhabit as if they were human may be overlooking the importance of this sort of nuanced colonization of these sites.

The Twitter bots discussed above have aspects in common with Japanese character bots, which Keiko Nishimura notes are also "not necessarily concerned with convincing others that they are human" (2017). Instead, character bots and, I would suggest the types of Twitter bots discussed above, demonstrate "the capacity of non-human action in online social media" (Nishimura, 2017). These bots become successful members of the social network not by passing as human, but rather by capitalizing on the playful and surprising ways that their non-humanness is emphasized as they interact with people and sometimes also with each other. It is the otherness of these bots—their overtly presented other-than-human alterity—that supports their social position on the network, as opposed to their ability to communicate and interact as if they were "human others."

Rethinking Social Robots as Alterities

Having introduced the potential to regard chatbots and Twitter bots as social, even when they declare themselves to be non-human and communicate in obviously non-humanlike ways, can the idea of what constitutes a physically instantiated social robot be similarly extended? It seems important to note up front that my washing machine, a relatively simple and decidedly non-humanlike device, although with sensors to judge the load and thus minimize water usage and washing time, nonetheless takes on a social role from my perspective due to the joy it affords me when it plays its completion song. Having assumed I would hate this "feature," I quickly discovered that the machine never fails to brighten my day when I hear it "sing." In common with many people, I have also developed attachments to a number of cars and computers in my time, but what about robots designed to interact with humans more directly? Can they be considered social even if they communicate in ways that are not overtly framed as humanlike?

Vyo is described as "an expressive social robot" that acts as an "embodied interface for smart-home control" (Luria et al., 2017, p. 581). This robot is overtly non-anthropomorphic, having been "inspired by a microscope metaphor" (p. 582). This design means that Vyo has a "body," "neck," and "head." When a person approaches Vyo, the robot raises its head to "face" them and "look" at them with its single camera "eye" (which resembles the viewing lens of a microscope). If the person moves on, the robot looks back down. Unless there is something wrong with the home, the robot will not try to engage with people; rather, it waits for a person to begin most interactions. To do this, the person chooses which system within the smart home to view and control by placing a physical icon, a "phicon," onto Vyo's "turntable" (which is effectively the microscope stage of the robot, to continue the design metaphor). The robot then bows its head to look at the "phicon," a movement which also reveals the small visual display on the back of its head. Vyo confirms recognition of the phicon, each of which represent a particular aspect of the smart home, by showing a matched icon on the screen. The person can control the home by manipulating the phicons. Placing a phicon on the turntable activates the related system; for example, turning the heating on. Moving the phicon up or down the turntable turns the thermostat up or down, the target temperature being displayed on the screen as it is altered.

Vyo's design is non-anthropomorphic, but the robot's behaviors nevertheless encourage anthropomorphism, such that humans read the robot's non-verbal cues through comparison with human face, head, and body movements. The design of this robot broadly sets aside touchscreen, voice, gestural, and augmented reality interfaces. Guy Hoffman, one of the robot's designers, does demonstrate the robot responding to voice commands with a voice of its own, but in the main communication with this robot is silent, embodied, and non-verbal (Vyo, 2016). Vyo's appearance invites people to interact with the robot as they would with a microscope, using it to gain more detailed information about an aspect of their domestic setting, but alongside this its communication is shown to rely on their ability to anthropomorphize its movements. Vyo is designed to seem "reliable, reassuring and trustworthy" (Luria et al., 2016, p. 1020). As mentioned above, when something is wrong in the home it uses "several peripheral gestures" to alert the user to situations. The first, "a nervous breathing gesture" of the head and neck, indicates that there is something that needs attention, but not urgently (p. 1023). This action is designed to cue someone seeing the robot to attend to the situation when there is time to stop what they are doing and check on Vyo. The second, "urgent panic," indicated by Vyo "looking around" with what seems to be an air of desperation is a more obvious movement that is less easy to ignore. This behavior is designed to encourage an immediate response from any person in the robot's vicinity (pp. 1023–1024).

The designers of this robot suggest that there is a division between those who think Internet of Things (IoT) devices should be autonomous and effectively invisible (i.e. requiring no human attention or interaction) and those who prefer the idea of "technology that promotes the user's sense of control and engagement" (p. 1019). In particular, they note that "Weiser envisioned the future of domestic technology to be calm, ubiquitous, autonomous, and transparent," whereas "Rogers … argues for a shift from proactive computing to proactive people" (p. 1019). Keeping this in mind, Vyo's design aims to be "engaging," yet also "unobtrusive." While Vyo is clearly "device-like," its behavior is also framed by its context as a robot that helps manage the smart-home environment in collaboration with the people who live there. This robot can be social (engaging, respectful and reassuring) when required, but also take a background position much of the time (unobtrusive, device-like). Vyo therefore bridges the division described above, by attempting to balance "autonomy and engagement" in one device (p. 1019). It is worth noting that, although Luria and colleagues suggest that all social robots have the capability to do this, many of them demand engagement far more strenuously than Vyo.

In contrast with Vyo, a robot designed as a platform to enable research into new types of interface for the smart home, Jibo is a commercial robot that has been designed for long-term home use. Crowd funding campaigns began for Jibo in 2014 but, following a series of development delays, Jibo has only just been shipped to early pre-order supporters. This robot is gendered, unlike Vyo, with the tagline for the Jibo website saying, "He can't wait to meet you." Jibo has a tapered cylinder as its "body" divided into sections allowing it to turn and sway on three axes. This body supports a hemispherical "head" section, the flat side of which contains a liquid crystal display. This display, Jibo's "face," most often shows a single circular "eye," which moves and changes shape supporting the sense of emotion in what the robot is saying, or the way its body is moving. Although this robot might be described as non-anthropomorphic, in a similar way to Vyo, Jibo's voice, associated body, and eye movements are designed to compel people to anthropomorphize the robot.

The promotional materials for Jibo suggest that this robot will eventually be able to complete many tasks, including being embedded into your smart-home environment. Jibo is situated as a personal assistant, taking voice messages, reading emails, reminding people of appointments, and reading stories with children (Jibo, 2015). Since Jibo's release, it has become clear that many of these skills are not yet available, although the robot's facial recognition and camera technology does mean it can take photographs of people when asked. Jibo's primary interface is vocal, and his voice and expressive "face" and body allow him to talk and emote in humanlike ways in conversation with people.

There are parallels between the design for Vyo and that of Jibo. Both of these robots have defined bodies, heads, and faces with eyes, but in spite of their similarities, their respective interactions with humans take very different paths. It is important to acknowledge that Vyo is a research robot, an experimental interface meant to allow its creators to test people's responses to something that breaks the normal mold of the social home robot. In contrast, Jibo has been launched as a commercial robot designed to become a companion and assistant for all members of the family. The goals of the roboticists working on these projects are very different from one other, but both these robots demonstrate how a robot can become part of a social network in the home. Vyo has been designed to become part of a person's domestic social network, but in a sufficiently unobtrusive way to "allow people to focus on developing close relationships within their family" (Luria et al., 2016, p. 1020). In contrast, Jibo's aim is to interact in ways that allow him to become part of the family.

Roberto Pieraccini, who worked on the prototype for Jibo, suggests that the design team's goal from the beginning was not to support the sense that Jibo was a living being, but rather "to remind people Jibo is in fact a robot" (interviewed in Rozenfeld, 2017). However, Jibo's non-anthropomorphic appearance and the machinelike tone to his voice may not be enough to achieve this goal. For example, Jeffrey van Camp notes that he began to feel "guilty" when he "left Jibo alone in the dark all day," a response that he links with the sense that Jibo was somewhat like a dog, but which may also have arisen alongside the way he and his wife "began to think of Jibo as a little person" (2017). Jibo coos when you pet his head, and shows "friendly curiosity in the way he leans back and looks up at you" (van Camp, 2017). Although Dave Gershgorn (2017) is not convinced that Jibo attains the level of being a "robotic friend," he nonetheless grants the robot the status of "robotic acquaintance," this decision seemingly driven by Jibo's technical limitations at his initial release. There is a sense in his statement that Jibo might become more of a companion as his software is upgraded over time. As Chris Davies also notes, many skills are "within Jibo's potential skill-set" (2017). Once ready these will be packaged alongside the robot's "cute" personality, meaning that in the future this robot could move from being a "curiosity" to "a useful little helper" (Davies, 2017). Geoffrey Fowler suggests, while people "have utilitarian relationships with most technology," robots such as Jibo "do things simply to elicit emotion" (2017). Interestingly, Vyo also elicits emotion, judging by my response to the video showing its "nervous breathing" and "urgent panic" behaviors (Vyo, 2016). For me, Vyo is certainly a social robot but, as Evan Ackerman emphasizes, Vyo is a "social robot that's totally, adorably different" (2016). Importantly, although this robot evokes an emotional response, it does not do this simply to draw people into engaging with its personality; instead, this robot has a clear and practically important aim, to communicate that something is wrong and requires human attention.

Theorizing Alterity

One way to theorize the very different interaction styles of Twitter bots and socialbots, or the social robots Vyo and Jibo, is by considering them through two concepts introduced by Sherry Turkle's ethnographic studies of human–technology relations. Her conception of "evocative objects," which "invite projection" (Turkle, 2005, p. 27), is particularly useful when considering Twitter bots and Vyo, both of which are communicative, but very clearly positioned as non-human and also relatively non-demanding in their social interactions. As I have already suggested, people may be amused or inspired to think by the utterances of Twitter bots and, while people can choose to engage with them directly, the bots described in this chapter rarely demand attention. The situation with Vyo is similar. The evocative nature of this social robot relies to some extent on its microscope-like appearance, with a social element introduced through its quiet, unobtrusive gestures and behaviors. Vyo only demands attention when absolutely necessary, and even then this is framed as a need for human assistance.

In comparison, socialbots and Jibo would seem to be better described as "relational artifacts," which "demand engagement" (Turkle et al., 2006, p. 315). Although not all socialbots demand responses, they are clearly positioned as conversational agents, whose seemingly humanlike nature requires a person to treat them as they might another human. While Jibo's physical presence means he cannot be confused so easily with a human communicator, his behavior is more insistent than that of Vyo. While Vyo looks up to show awareness when someone

walks past, Jibo actively looks for people and is constantly listening out for someone saying the phrase "Hey, Jibo." Although Jibo's visual and auditory surveillance
of the home is a necessary part of his successful operation, for some people his
endless curiosity and watchful eye might become unsettling as they realize he
"won't stop staring" at them (van Camp, 2017). Jibo demands engagement from
people whenever possible, although he can be told to go to sleep if his inquisitive
nature becomes too much to bear.

These two ways to interpret bots and robots, as evocative objects or relational
artifacts, highlight their very different types of personality. An evocative object
reveals itself gradually through its communication and behavior. Such a robot may
encourage anthropomorphism, but it does so without making people assume it
should be capable of a humanlike reciprocal response. In contrast, the relational
artifact, with its intense focus on direct engagement is more likely not only to be
considered humanlike, but also to be compared against a human helper or companion. Positioning a bot or robot as evocative may therefore mean the machine
is less likely to be judged as deficient in comparison with a human ideal, as well
as allowing it more freedom to communicate and interact in novel, non-human
ways.

An alternative, more philosophical perspective from which to theorize relations with robots can be developed by extending Emmanuel Levinas's conception of "the face to face" that occurs during human–human encounters (1969).
Within such an encounter the human other reveals themselves through what
Levinas terms the "face" (1969, pp. 79–81). Importantly, this is not a physical face
and its expression, but rather a more transcendent property of the other, which
encapsulates all the ways the other can express their personality or give a sense of
their being. Levinas argues that, while it is not possible to "entirely refuse the face
of an animal," the face they reveal "is not in its purest form" (Wright et al., 1988,
p. 169). Furthermore, as one might expect, he is also skeptical that objects could
ever reveal a face (Levinas, 1989, p. 128). However, a number of scholars have
argued that both animals (Derrida, 2002; Clark, 1997) and robots (Gunkel, 2012;
Sandry, 2015) can reveal a "face."

In relation to this chapter, there are two reasons for extending Levinas's theory
to consider human–robot interactions. The first is that his primary concern is to
retain the absolute alterity of the other, even as self and other are drawn into the
proximity of an encounter, a factor that is particularly important when considering the non-human agency of machines. It is worth noting that for Levinas proximity does not imply physical closeness, but rather is part of the act of revealing
the self through a face. Thus, even online, self and other can be drawn into proximity through seeing each other's communications or through direct interaction.

The second is that Levinas's face to face requires the human self to respond
to the other, in this case a machine other, without the expectation of reciprocal
action. In extending Levinas's thought to consider social machines, it is important
to stress that the self's response need not involve treating the machine other as if it

were human, but rather the human should respond to and respect the machine for its particular non-human abilities in interaction. Although machines may respond to humans, as is the case with some Twitter bots, Japanese character bots, Vyo and Jibo, these non-human others are unlikely to reciprocate on equal terms; however, Levinas's conception of the face to face does not require an interaction to be reciprocal. Indeed, as Amit Pinchevski explains, Levinas's "provocative specula-tion" is that the face to face "is asymmetrical" (2005, p. 9).

Extending Levinas's conception of the face to face to consider human–bot and human–robot encounters, such as those discussed in this chapter, is therefore valuable because it emphasizes the continual presence of otherness, in this case a machine otherness. It also allows space for bots and robots to communicate and respond in their own particular ways, with no need for reciprocal action—which also means, no need to be overtly humanlike—in their communication and behavior. In addition, Levinas's theory as it originally pertains to human–human encounters serves as a strong reminder of the differences that are always present even between human individuals. This brings the chapter full circle, to suggest that designs for socialbots and robots that idealize human modes of communication and interaction, whether online or in physical spaces, should instead attend to the many different ways in which humans communicate within their social networks.

Conclusion

I am suspicious of the idea that it is important to "humanize technology," the ulti-mate goal that Breazeal mentions in her introduction to Jibo (Jibo, 2015), in par-ticular when this is linked with the creation of interfaces that support a sense of "ideal" humanlike communication, when many humans may not communicate in this "idealized" way themselves. From a practical perspective, creating social robots that do not rely heavily upon one form of communication and interface may be important. As Hoffman notes, the physically embodied interface of Vyo may be more "democratic," offering a new way for "populations who would have a harder time navigating a complex on-screen menu or a voice interface" to interact with a social robot (cited by Ackerman, 2016). Innovative and mixed interfaces may have the potential to support people with disabilities, not to exclude them from using new technologies that have the potential to help them in their everyday lives.

This chapter suggests that there is something to be gained from expanding the types of machine, interaction style, and communication that are accepted as social beyond what seems to be an assumed, but not often critically considered, sense of ideal humanlike communication and sociality, whether this plays out within digital spaces or in the context of face-to-face interaction. As Mavridis suggests, "the space of potentialities for artificial agents within social networks is quite vast," in relation to their appearance (in terms of embodiment or avatar), whether they declare themselves as machines (if this is not clear), their level of autonomy (which may shift) and the perceived personality they convey through their actions (2011, pp. 298–299). Rather than considering how technology might augment human selves directly, this chapter is therefore more concerned with the way that communicating with diverse machines might highlight the need to recognize and respect all sorts of others as part of human social networks, even when these others do not, or cannot, follow an idealized mode of human communication, expression, and social engagement.

Notes

1 The @gardenBritish bot was created by Thomas McMullan (@thomas_mac). The @MagicRealismBot was created by Chris Rodley (@chrisrodley) and Ali Rodley (@yeldora_).
2 The @tinycarebot was created by Jonny Sun (@jonny_sun).
3 The @softlandscapes bot was created by George Buckenham (@v21) and the @dailyasteri bot by @jordanphulet.

References

Ackerman, E. (2016). Vyo is a fascinating and unique take on social domestic robots. *IEEE Spectrum*, June 9. https://spectrum.ieee.org/automaton/robotics/home-robots/vyo-robotic-smart-home-assistant

Boshmaf, Y., Muslukhov, I., Beznosov, K., & Ripeanu, M. (2011). The socialbot network: When bots socialize for fame and money. *Proceedings of the 27th annual computer security applications conference* (pp. 93–102). New York: ACM.

boyd, d. (2014). *It's complicated: The social lives of networked teens.* New Haven: Yale University Press.

Breazeal, C. L. (2002). *Designing sociable robots.* Cambridge, MA: MIT Press.

Clark, D. (1997). On being "the last Kantian in Nazi Germany": Dwelling with animals after Levinas. In J. Ham & M. Senior (Eds.), *Animal acts: Configuring the humans in Western history* (pp. 165–198). New York: Routledge.

Compton, K. (n.d.). Opulent Artificial Intelligence: a manifesto. http://www.galaxykate.com/pdfs/galaxykate-zine-opulentai.pdf

Davies, C. (2017). Jibo Review: Alexa gets some cute competition. *Slash Gear*, November 13. www.slashgear.com/jibo-review-2017-13507668/

Derrida, J. (2002). The animal that therefore I am (more to follow). *Critical Inquiry*, 28(2), 369–418.

Donath, J. (2007). Signals in social supernets. *Journal of Computer-Mediated Communication*, 13(1), 231–251. doi: 10.1111/j.1083–6101.2007.00394.x

Dunbar, R. I. M. (2004). Gossip in evolutionary perspective. *Review of General Psychology*, 8(2), 100–110. doi: 10.1037/1089–2680.8.2.100

Feldman, J. (2016). The bot politic. *The New Yorker*, December 31. www.newyorker.com/tech/elements/the-bot-politic

Flatow, I. (2016). *I, Twitter bot*, August 19. www.sciencefriday.com/segments/i-Twitter-bot/

Fowler, G. A. (2017). These robots don't want your job. They want your love. *The Washington Post*, November 17. www.washingtonpost.com/news/the-switch/wp/2017/11/17/these-robots-dont-want-your-job-they-want-your-love/?utm_term=.37502e737650

Gehl, R. W., & Bakardjieva, M. (2017). Socialbots and their friends. In R. W. Gehl & M. Bakardjieva (Eds.), *Socialbots and their friends: Digital media and the automation of sociality* [Kindle edition]. New York: Routledge, Taylor & Francis.

Gershgorn, D. (2017). My long-awaited robot friend made me wonder what it means to live at all. *Quartz*, November 8. https://qz.com/1122563/my-long-awaited-robot-friend-made-me-wonder-what-it-means-to-live-at-all/

Gunkel, D. J. (2012). *The machine question: Critical perspectives on AI, robots, and ethics.* Cambridge, MA: MIT Press.

Jibo. (2015). *Jibo: the world's first social robot for the home.* Video. www.youtube.com/watch?v=3N1Q8oFpX1Y

Jones, S. (2015). How I learned to stop worrying and love the bots. *Social Media + Society,* 1(1). doi: 10.1177/2056305115580344

Levinas, E. (1969). *Totality and infinity.* Pittsburgh: Duquesne University Press.

Levinas, E. (1989). Is ontology fundamental? *Philosophy Today,* 33(2), 121–129.

Luria, M., Hoffman, G., Megidish, B., Zuckerman, O., & Park, S. (2016). Designing Vyo, a robotic smart home assistant: Bridging the gap between device and social agent. *IEEE International Symposium on Robot and Human Interactive Communication.* New York: IEEE Press.

Luria, M., Hoffman, G., & Zuckerman, O. (2017). Comparing social robot, screen and voice interfaces for smart-home control. *Proceedings of the 2017 CHI Conference on Human Factors in Computing Systems* (pp. 580–628). New York: ACM Press. doi: 10.1145/3025453.3025786

Mavridis, N. (2011). Artificial agents entering social networks. In Z. Papacharissi (Ed.), *A networked self: Identity, community and culture on social network sites* (pp. 291–303). New York: Routledge.

Mavridis, N., Datta, C., Emami, S., Tanoto, P., Ben-AbdelKader, C., & Rabie, T. F. (2009). Facebots: Social robots utilizing and publishing social information in Facebook. *Proceedings of the IEEE Human-Robot Interaction Conference.* New York: IEEE Press.

Mitchell, N. (2017). Alexa, Siri, Cortana: Our virtual assistants say a lot about sexism. *ABC News,* August 11. www.abc.net.au/news/2017-08-11/why-are-all-virtual-assistants-female-and-are-they-discriminatory/8784588

Nishimura, R. W. (2017). Semi-autonomous fan fiction: Japanese character bots and non-human affect. In R. W. Gehl & M. Bakardjieva (Eds.), *Socialbots and their friends: Digital media and the automation of sociality* [Kindle edition]. New York: Taylor & Francis.

Pinchevski, A. (2005). *By way of interruption: Levinas and the ethics of communication.* Pittsburgh: Duquesne University Press.

Rozenfeld, M. (2017). Jibo: The friendly robot that makes you feel at home. *The Institute, IEEE,* April 5. http://theinstitute.ieee.org/technology-topics/artificial-intelligence/jibo-the-friendly-robot-that-makes-you-feel-at-home

Sandry, E. (2015). *Robots and communication.* New York: Palgrave Macmillan.

Turkle, S. (2005). *The second self: Computers and the human spirit.* Cambridge, MA: MIT Press.

Turkle, S., Breazeal, C., Dasté, O., & Scassellati, B. (2006). First encounters with Kismet and Cog. In P. Messaris (Ed.), *Digital media: Transformations in human communication* (pp. 303–330). New York: Peter Lang.

van Camp, J. (2017). Review: Jibo social robot. *Wired,* November 7. www.wired.com/2017/11/review-jibo-social-robot/

Vyo (2016). *Vyo—Social robot for the smart home—scenario walkthrough.* Video. www.youtube.com/watch?v=pi3fAyNyClw

Wright, T., Hughes, P., Ainley, A., Bernasconi, R., & Wood, D. (1988). The paradox of mortality: An interview with Emmanuel Levinas. In R. Bernasconi & D. Wood (Eds.), *The provocation of Levinas: rethinking the other* (pp. 168–180). London and New York: Routledge.

7

BEYOND EXTRAORDINARY: THEORIZING ARTIFICIAL INTELLIGENCE AND THE SELF IN DAILY LIFE

Andrea L. Guzman

Artificial intelligence is extraordinary—at least, that is how it has been culturally conceived. AI with its goal of imbuing machines with humanlike intelligence has been portrayed as the pursuit of a "dream" (Ekbia, 2008) or, even more extraordinarily, described as humans stepping into the role of a god (e.g. McCorduck et al., 1977). Similarly, the implications of AI for self and society have been cast at a level well beyond that of ordinary technology. The promises and consequences of AI have not been so much about who we are as individuals; rather, at stake is who we are as humans and what it means to be human (e.g. Bostrom, 2016; Haugeland, 1985; Turing, 1950; Turkle, 1984). AI seemingly places our collective humanity up for grabs in ways that can elevate or decimate the human race, depending on how it is implemented (e.g. Leverhulme Centre for the Future of Intelligence, 2017).

For these reasons, I found myself perplexed a few years ago while conducting a research project regarding AI. The study's focus was to better understand people's conceptualizations of vocal social agents (VSAs)[1]—AI-enabled, voice-based technologies such as Apple's Siri—and how these perceptions played back into people's own understanding of the self in relation to AI (Guzman, 2015). But, I did not encounter people questioning human intelligence or their own abilities after chatting with Cortana. There was no wringing of hands or zealous optimism regarding what these programs meant for them as individuals, let alone humanity, now or in the future. Instead, people's reflections on the self in light of agents contained themes similar to those associated with everyday technologies, such as the mobile phone. People thought of themselves as lazy for vocally asking an agent for directions instead of just typing an address into the device. They did not want people to perceive them as being rude for talking to their phone while around others. Users were more interested in how a sassy Siri could make them the center of attention than in what a disembodied voice complete with personality meant for the future of the human race. Neither the technology nor its impact on the self from the perspective of users seemed extraordinary; rather, the self in relation to talking AI seemed to be, well, ordinary—just like any other technology.

Why such a disconnect between what scholars have long touted as the implications of AI for the self and what participants were telling me? The answer could be technological. Cortana and company routinely fail at their job, reminding us that they are nowhere near human. VSAs do not possess the full range of human-level intelligence and have limited application. They are not taking jobs or even performing tasks outside the context of a personal assistant. The answer also could be methodological. Maybe I was not asking the right questions or taking into account other ways in which aspects of the self can manifest within conversations about AI and technology. The self and AI are not exactly subjects people routinely consider or talk about.

I have come to realize, however, that my struggle to make sense of my data was a theoretical problem that had intertwined technological and methodological components. I had grounded my research in the predominant conceptualization of AI as extraordinary: AI has been portrayed as standing apart from other technological endeavors because it seeks to recreate the human mind within the machine. By its very nature, then, AI has been said to challenge our own human nature (e.g. Dechert, 1966; Haugeland, 1985; Turing, 1950; Turkle, 1984). The result has been that for most of the history of AI, the self in relation to AI has been theorized primarily at the metaphysical level, with emphasis placed on the implications of machine intelligence for our humanity. However, in my research, I was interested in studying the self at the individual level, as a way of theorizing back to the metaphysical level, not vice-versa. Furthermore, I was looking at a specific AI technology, not AI as a whole. My study also was taking place almost 70 years removed from when AI and its implications for humanity were first theorized. In the interim, particularly within the past decade, artificial intelligence has evolved significantly. In short, the predominant way of conceptualizing AI and theorizing AI in relation to the self could no longer account for what AI had become. The old theory did not fit the new reality.

The purpose of this chapter is to further challenge the predominant ways we have theorized AI in relation to the self while providing a new approach to this research that is better attuned to the evolving nature of artificial intelligence. I have specifically written this chapter for scholars like myself who study AI but are outside the field and face the difficult task of adapting and merging existing theories of AI with theories in our own areas. To that end, I begin by providing a brief overview of the historical, technological, and theoretical context in which the self came to be theorized in terms of AI's inherent disruption to human nature. Next, I explain ongoing and emerging problems with this predominant view of AI and the self. Once readers have had a chance to understand why and how existing theories of AI and the self are inadequate for the study of today's AI, I introduce a research approach that addresses these shortcomings by drawing from emerging theory in human–machine communication.

AI as Extraordinary: The Stakes for Humanity

The first group to suggest that artificial intelligence would pose a challenge to human nature were the very scholars working to make AI a reality; although, their motivation for doing so varied. For Turing, broaching people's reactions to the humanlike qualities of thinking machines was a pragmatic issue (Gandy, 1999). He anticipated that categorizing machines as intelligent would disrupt long-held beliefs about human nature, and so when writing about machine intelligence, Turing (1948, 1950) confronted these objections head-on, refuting them. Where Turing saw a potential hurdle to scientific advancement, others saw an opportunity. Some scholars tapped into the human desire to control nature, including their own, as a way to promote AI. McCorduck and colleagues describe the pursuit of AI as nothing short of a means through which humans may triumph over nature, adding "Artificial intelligence comes blessed with one of the richest and most diverting histories in science because it addresses itself to something so profound and pervasive in the human spirit" (1977, p. 954). Other scholars warned that such challenges to human nature would upend people's sense of their place in the social and natural order. In advocating for scholars to consider the consequences of automation, Theobald (1966) declares, "Man will no longer need to toil: he must find a new role in the cybernetics era which must emerge from a new goal of self-fulfillment" (pp. 68–69). Since AI's founding, the implications of AI for our humanity have been an integral part of AI research.

Any interaction with technology has the potential to influence the self, and, in that regard, AI is not unique. However, volumes have not been written about how garage door openers will force us to rethink our humanity. Even technologies that have garnered significant scholarly attention regarding the self, such as social media (e.g. Papacharissi, 2011), have not generated a level of metaphysical inquiry anywhere near that of AI. Underpinning the pursuit of AI has been the theoretical conceptualization of the mind as machine (Haugeland, 1985). What has historically situated AI as extraordinary, setting it apart from garage doors, social media, and other technologies is that AI seeks to recreate a part of us that we have conceived as being uniquely human—our minds. It is our minds that separate us from animals and objects and are key to our humanity, or so we have thought. As Turkle (1984) explains of AI scholars: "In the course of exercising their profession, they have made questions about human intelligence and human essence their stock and trade" (p. 20). AI's challenge to humanity has been promulgated in the ways in which researchers have theorized AI (Ensmenger, 2012) as well as in the scientific and cultural discourse surrounding AI (Ekbia, 2008).

The pitting of human against machine in some sort of intellectual competition has been an integral part of the development and testing of AI theory and technology (Ensmenger, 2012) and in setting the stakes for AI and the self. The precedent of theorizing AI through intellectual face-offs between human and machine begins with Turing's (1950) "test" that set the bar for machine intelligence with a computer program's ability to pass itself off as a human typist. Work toward developing automated chess programs that could beat human opponents also became an important testing ground for AI theory and technology (Ensmenger, 2012). The use of chess was significant at the time because chess has a long history of being conceptualized as *the* game requiring the utmost level of human intellect (Ensmenger, 2012). In using chess as the field's theoretical testbed, AI scholars raised the ontological stakes for humans and machines to one of its highest levels, if not *the* highest level at the time. Computers could transcend their nature by beating humans at their "best" game, while humans were defending their unique ontological position. When IBM's Deep Blue finally beat the world's best chess player, the moment was hailed as a triumph for machines and a loss for humanity (Bloomfield & Vurdubakis, 2008). This scenario has played out repeatedly as new intellectual battles between human and machine—such as a game of Jeopardy (Kroeker, 2011) or, more recently, Go (Anthes, 2017)—have continued to push the stakes of AI to higher and higher levels.

Our understanding of AI and its meaning for the self also are established through what Ekbia (2008) calls the "talk of AI." This talk includes how scientists explain the concept of AI, frame its issues, and discuss its findings as well as media representations of AI, including science-fiction plot lines. The idea of thinking machines and the connection between humans and machines has garnered significant media interest and speculation. According to Ekbia (2008) one of the reasons that AI has attracted such media attention is "that people sense that the issues raised by AI are directly related to their self-image as human beings" (p. 319). It is of little surprise, then, that the talk of AI regarding the self remains fixated on our human nature. The theory of mind as machine, around which AI was founded, is itself a metaphor, as is Simon's (1999) later inversion, "machine as mind." In both metaphors, the nature of the (human) mind and the nature of the machine are being compared for the sake of making a conceptual connection between the two (as well as what AI researchers hoped would eventually be a theoretical and technological one). In addition, AI technologies and their function are primarily described in human terms (Ekbia, 2008). The heavy use of metaphor and anthropomorphic language in the talk of AI reflects the theoretical claims of the connection between human and machine underpinning AI while, simultaneously, reinforcing them (Agre, 1997; Ekbia, 2008). In doing so, the talk of AI also perpetuates a focus on AI's implications for our humanity.

The factors I have discussed so far—the theories underpinning AI, the methods of studying AI, and the talk of AI—are integral to forming and maintaining a focus on the meaning of AI for our human nature. However, there is one metafactor underlying all these others that needs to be discussed: the origin of AI in speculation based on technological reality. AI did not begin with the creation of a mind within a machine. Rather, it begins with an idea of the future potential of computers. Today, computers are taken for granted, but in the mid-twentieth century, their capabilities were mind-boggling. It is important to remember that it was Turing who played an integral role in the development of computing at this time and, thus, was one of the few people in the world who had access to these machines and had witnessed their capabilities. As Gandy (1999) has argued, Turing's (1950) famous essay on machine intelligence is better understood as "propaganda" aimed at convincing people to realize that "computers were not merely calculating engines but were capable of behavior which must be accounted as intelligent" (p. 123). For Turing, computers already had a degree of intelligence; therefore, it followed that they could be made to be more intelligent. Similarly, the 1956 Dartmouth workshop that is considered the beginning of formal AI research within the United States originated out of an idea of what could be possible with computers. Its proposal reads, "The study is to proceed on the basis of the conjecture that every aspect of learning or any other feature of intelligence can in principle be so precisely described that a machine can be made to simulate it" (McCarthy et al., 2006, p. 12). The result of this "conjecture" was a field formed entirely around an idea of what "could be." Graubard (1988) explains the power of speculation in shaping the unique character of AI research:

> Had the term artificial intelligence never been created, with its implication that a machine might soon be able to replicate the intelligence of a human brain, there would have been less incentive to create a research enterprise of truly mythic proportions. In a very fundamental sense, AI is something of a myth.
>
> *(p. v)*

Myth here carries the denotative meaning with a positive connotation of the pursuit of something of great consequence that has yet to be achieved.

Conceptualizations of who we are in relation to AI, then, have formed around the myth that is AI. The pursuit of a myth meant that the daily work of AI and the technologies developed were interpreted and evaluated not only within the context of how they furthered understanding of the mind and technology in the present but also based on how they contributed to the ultimate goal of AI that was yet to be realized. And so, advances in AI have served as a way of thinking about what could be possible in the future. This is one of the many reasons why AI has been not only a technological pursuit but also a philosophical endeavor (Dennett, 1988). As Dennett (1988) explains, "Most AI projects are explorations of ways things might be done and as such are more like thought experiments than empirical experiments" (p. 289). These thought experiments extend to the implications of AI for who we are. Analysis of the self in AI has taken place within the context of the future, one in which the ultimate goal of AI—which presents the highest stakes for humanity—is realized.

Reconciling the Future of AI against Its Present

The predominant way we have theorized AI in relation to the self has remained fixated on what the future realization of AI would mean for our humanity.

However, problems exist with this theoretical approach to AI and the self. The idea and study of artificial intelligence is not without its critics, and some of the key aspects underlying the study of AI that also serve as the basis for how we theorize the self have been called into question. In addition, the discourse of AI that perpetuates the connection between human and machine, and thus reinforces the stakes of AI for humanity, does not always accurately represent the technological reality of AI. Furthermore, artificial intelligence—the concept, its study, and the resulting technologies—has continued to evolve while the predominant theory of the self in AI has largely remained static. The current state of AI and its integration into our daily lives cannot be fully understood through a theoretical lens developed around mid-twentieth-century technology and the possibilities of the future.

Metaphors drawing parallels between humans and machines have figured heavily in AI theory and discourse (Agre, 1997; Ekbia, 2008; Haugeland, 1985; Neisser, 1966). Some scholars have expressed concern with the central role of these metaphors in how we conceptualize and talk about AI (e.g. Agre, 1997; Dreyfus, 1999; Neisser, 1966). Their criticism is not so much in the general use of metaphor in AI research (see Agre, 1997); rather, scholars have been critical of how the metaphors connecting body and machine have been so thoroughly subsumed that they are no longer questioned (e.g. Neisser, 1966) and, in some cases, have stopped functioning as metaphors altogether (e.g. Agre, 1997; Ekbia, 2008). As metaphor, the meaning of "body as machine" draws a comparison between humans and machines, but when the comparison is no longer approached as a metaphor, the body becomes a type of machine. Within some veins of AI research the literal application of the metaphor has replaced the figurative (Haugeland, 1985). These same metaphors also have played an integral role in theorizing AI as intrinsically challenging our human nature. As the figurative aspect has receded, the relationship between human and machine is no longer one of comparison; instead, they are equated to one another. The ontological line that once stood between human and machine (Turkle, 1984) is erased, and, thus, the challenge of AI to human nature has seemingly been firmly established. The result is not only a potentially erroneous way of viewing the self in relation to AI but also a quashing of the need to look for any alternative ways of theorizing the self and AI.

Researchers also have questioned the framing of AI's capabilities within the context of human traits. They argue that anthropomorphic portrayals of AI technology can be false or, at the very least, misleading because they focus on the similarities between human and machine without accounting for the differences (e.g. Agre, 1997; Ekbia, 2008). The high-stakes "intellectual battles" between human and machine can also misrepresent advances in AI and just how close AI is coming to achieving human-level intelligence. While the creation of a chess program that could beat humans seemed to be a sign of success for AI, Ensmenger (2012) explains that the technological reality was more complicated. Yes, chess programs could play chess, and play it well, but the technology behind these programs contributed less to the advancement of AI than work in other areas. IBM's Watson (Kroeker, 2011) and Google's AlphaGo (Anthes, 2017) do have impressive AI capabilities and applications for AI well beyond the games they play against humans. Still, as with chess, the battles between humans and these technologies are largely symbolic, in that they do not represent the complexity with which humans and AI compare and contrast to one another. Overall, AI theory and the talk of AI do not just focus our attention toward one aspect of understanding the self—the implications of AI for humanity. They often create and reinforce a skewed picture of AI and its meaning for the self.

Like any physical portrait, this picture of AI has largely remained frozen, capturing what AI was thought to be as it was first emerging. Early efforts in AI coalesced around the creation of technology possessing general intelligence (Franklin, 2014). Now called artificial general intelligence, this is the conceptualization of AI as extraordinary that is theorized as putting our human nature up for grabs. But whether this level of AI can be achieved has also been the subject of debate (Dreyfus, 1999; IEEE, 2017b). Recent advances in technology, new ways of theorizing knowledge, and a better understanding of the human brain have given some additional credence to the feasibility of realizing human-level AI (Bostrom, 2016; IEEE, 2017a; Müller, 2016). Some scholars have even gone as far as to consider the point when technology surpasses human intelligence (e.g. Bostrom, 2016; Kurzweil, 2005). As with human-level AI, the stakes for machine "superintelligence" are high, according to Bostrom: "This is quite possibly the most important and most daunting challenge humanity has ever faced. And—whether we succeed or fail—it is probably the last challenge we will ever face" (2016, p. v). Although aged, the portrait of AI formed in its early years has staying power.

But, as myth, AI also has nebulous aspects, so that when viewed from a different angle or without its prominent frame, the picture of AI can shift and so too can its meaning for the self. Artificial intelligence grew out of a set of ideas regarding what could be possible with computers, but these ideas were never codified into a universal definition of AI (Franklin, 2014). The Dartmouth workshop that established AI research within the United States did not operationalize AI, referring to it only as the "artificial intelligence problem" (McCarthy et al., 2006). Moor (2006) explains: "The field of AI was launched not by agreement on methodology or choice of problems or general theory, but by the shared vision that computers can be made to perform intelligent tasks" (p. 87). That "vision" did initially center around the pursuit of artificial general intelligence as a means to better understand the human mind, but even that work followed divergent theoretical and methodological paths (Dreyfus & Dreyfus, 1988). Eventually the intertwined goals of understanding the mind through building technology became two research trajectories within AI: the study of the mind and the engineering of "intelligent" machines (Franklin, 2014). These trajectories can intersect, but, more often, proceed separately (Boden, 2016), drawing from and contributing back to varied conceptualizations of AI.

Work within the engineering of AI, in particular, has offered an alternative way of theorizing AI. As the field progressed, some scholars shifted their focus away from the long-term goal of creating a computerized mind toward the shorter-term objective of the commercialization of technology that could carry out certain tasks requiring a specific type of knowledge. Forty years after AI's founding, Winston (1987) summarized the field's different priorities: "The primary goal of Artificial Intelligence is to make machines smarter. The secondary goals of Artificial Intelligence are to understand what intelligence is (the Nobel laureate purpose) and to make machines more useful ("the entrepreneurial purpose" (p. 1). While AI scholars initially sought to create autonomous machines to understand the human mind, researchers began to view the building of intelligent devices as an end unto itself. Today, the engineering of smart devices is the prominent area of research within AI (Boden, 2016). While efforts toward artificial general intelligence are ongoing, most of the AI programs found in commercial applications are categorized as narrow AI (Holdren & Smith, 2016), technology that operates "intelligently within some relatively narrow domain" (Franklin, 2014, p. 16).

Advances in AI as well as in computing generally also have further pushed AI out of research labs and into everyday spaces. Boden (2016) explains just how widespread AI has become: "AI's practical applications are found in the home, the car (and the driverless car), the office, the bank, the hospital, the sky ... and the Internet, including the Internet of Things (which connects the ever-multiplying physical sensors in our gadgets, clothes, and environments" (p. 1). The AI that we daily interact with—whether or not we realize it—are neither the fanciful sentient entities of science fiction nor the higher-intelligence technologies some AI researchers aspire to build. Instead, we encounter AI technologies that perform specific tasks throughout the various aspects of our lives. Yet the predominant lens available to us for theorizing the self within the context of AI was formed around the AI technologies of a fictional tomorrow, not today.

Returning to my opening anecdote, it is now easy to see why I was so baffled by my data. I had developed my original interview questionnaire grounded in the predominant view of the self in relation to AI, with AI being conceptualized within the context of artificial general intelligence; however, I was studying the self in relation to a very different type of AI (a voice-based agent) that operates within a specific framework, as an assistant, and is encountered within the mundane aspects of everyday life, not an experience projected onto the future. I was applying a theory of the self that was developed around AI as extraordinary to the study of AI that was situated within the ordinary. The theoretical lens was an inadequate match for my subject.

AI within the Ordinary

Until recently, AI has been theorized as extraordinary and primarily existed outside the realm of the ordinary. AI has been physically restricted to research labs, theoretically projected onto the future, and culturally confined to the media, meaning that the average person had little-to-no opportunity to experience AI first hand. Therefore, it also has been difficult for scholars to study the implications of AI for the self at the individual level (another reason why theory regarding the self in AI remained stagnant). The evolution of AI that necessitates a new way of theorizing AI and the self has also loosened some of these former barriers to this research. Scholars have the opportunity to study how individuals conceptualize themselves in relation to AI that is increasingly part of their everyday lives. The challenge now is finding a theoretical and methodological way forward that is not restricted to what AI once was or was projected to be but that is responsive to what AI is and who we are.

Theory being developed in the emerging area of human–machine communication provides one such approach. In my research regarding artificial intelligence (Guzman, 2015, 2017) and technology more generally (Guzman, 2016, In Press), I have advocated for scholars to conceptualize interactions with devices and programs as a form of communication. From this theoretical perspective, a person does not simply use AI, as one would use a tool; rather, they make sense of AI in a dynamic process *with* these technologies. Both the human and the machine play an active role in how the machine is understood (see also Neff & Nagy, 2016). The technology, in this case AI, is no longer positioned as only a medium, passing information among humans, but is also theorized as functioning as a type of communicator.

The utility of this theoretical conceptualization of our interactions with AI is that it opens up a way to simultaneously understand how we view a specific AI technology and how we see ourselves in light of it. The process of communication is at the heart of how we come to know others and ourselves (Blumer, 1969; Goffman, 1959; Mead, 1934); this includes both direct interaction with a communication partner and indirect communication with others about them. As I have adapted this epistemological position originally formed around humans to a human–machine context, the machine becomes a communication partner. The focus then is on how we make sense of the AI technology through what unfolds in our interactions with it as well as what we learn about it from others, and how we then assess who we are in relation to a specific AI program or device. Intrinsic in this phenomenological approach to human–machine communication is that the conceptualizations of self and AI being formed are understood as temporal and contextual.

Following this approach, the study of who we are in relation to AI involves taking into consideration both the characteristics of the particular technology in question and the various elements that contribute to the self. As I tried to make apparent earlier in this chapter, simply referring to a technology as AI is not sufficient. Given the ambiguity surrounding the different types of AI, it is incumbent upon scholars to understand what type of AI they are studying and how its characteristics may potentially affect how people understand the technology and themselves. An in-depth example of assessing an individual AI technology from a communication perspective is available in Guzman (2017), so here I touch on only a few key elements.

Given the different ways AI is conceptualized, scholars should determine where a particular technology fits within AI as a whole: Who built the technology? What is its purpose? What does the technology do? What is the level of intelligence involved? Is the technology geared toward a highly specific (narrow) task or does it operate more generally? Attention also should be focused on aspects of people's direct interactions with the technology that inform their understanding of it: In what context(s) do people interact with or encounter the technology? How does the interface present the technology to the user? How does interaction take place between someone and the technology, including the modes of communication and the messages (verbal and non-verbal) exchanged? Because what we know of others is also built through what we are told about them, researchers may need to take into account the messages regarding a particular technology and AI more generally: How does the company market the technology? How do media depictions frame the technology? What similarities exist between the technology being studied and pop cultural depictions of AI?

Beyond their own assessment of the AI technology, scholars will have to account for users' own interpretations of the technology. As I learned in my research regarding vocal social agents, users have wide-ranging definitions of what constitutes AI. Some people thought of Cortana, Siri, and Google's voice-technology as AI because the programs could answer questions and provide information fairly well. Other people thought agents were a weaker form of AI or something that was close to but not quite AI, because the agents had flaws or did not possess the full-range of human intelligence. Still others did not consider agents to be AI because the programs were too simple to be considered intelligent, and a few people could not categorize these technologies because they did not know enough about AI to make a determination. In deciding whether Siri and similar programs were AI, people drew on knowledge about AI gained from personal or professional experience as well as media representations. They also compared and contrasted AI against other technologies. In some cases, people defined AI in similar ways but reached different conclusions as to whether agents qualified as AI.

Users' wide-ranging conceptualizations of AI further underscore the complexity of the study of AI and the self. If a person does not consider a device or program to be AI, when from a technological and research stand point it could be defined as AI, how are we to interpret their understanding of self in relation to AI? What if a participant's conceptualization of AI is based on a different but also accepted definition of AI? It is entirely possible that in a research project a participant and the researcher may hold divergent but externally valid definitions of AI. Approaches like the one I have outlined here dictate that careful attention be paid to the etic and the emic and to the weighing of one against the other. In that, the study of AI is no different. But AI's nebulousness and relative newness, as an actual versus fictional technology, require extra care in formulating our research questions and in interpreting the data.

This is why it is so important for us as researchers to guard against our own fixation with the extraordinariness of AI that has been so entrenched within the predominant ways that AI has been theorized (and why I dedicated two-thirds of this chapter to explaining why this view is so pervasive and problematic). What led me to the theoretical breakthrough underlying this chapter was a decision to look beyond the dominant view of AI in relation to the self. Rather than focus on people's reactions to and interpretations of the human-like traits of agents—the characteristics of AI that are supposed to challenge our human nature—I shifted my focus to understanding what my participants told me again and again mattered the most to them, the "ordinary" aspects of AI as technology. People routinely compared agents to everyday technologies (mobile phones, navigation devices) and expectations surrounding technology generally (utility, usability, etc.) to decide whether, how, and when to use an agent. This is not to say that people did not recognize humanlike elements within these agents. These characteristics—such as the ability to speak, to learn, and to enact a particular personality—played critical roles in how people made sense of agents. But most people viewed these humanlike traits within the context of the machine, assessing whether these traits made agents a better technology than other programs and devices.

The technological aspects of voice-based agents also mattered the most for how people incorporated these technologies into performances and perceptions of the self. Some people who adopted Cortana or Siri regularly used the agents because they viewed themselves as being tech-forward. Others rejected agents because they considered the programs to be yet another example of unnecessary technology and prided themselves on not falling into these trends or becoming overly dependent on technology. Relatedly, some users described themselves or other people as being lazy for using a voice-based technology when typing a request worked well enough. And so, a limited level of intelligence and the ability to communicate did not set up an ontological face-off between Cortana, a machine, and the user, a human. Nor did the humanlike elements of agents invoke in most people an existential reflection on what it means to be human. In the minds of people using vocal social agents, what is at stake is their productivity, not their humanity.

Beyond Extraordinary: AI as Ordinary

As I was writing this chapter, the professional organization IEEE released a special magazine issue on AI (IEEE, 2017a) with a Q&A feature asking prominent AI scholars to weigh in on the future of artificial general intelligence. When asked "how brainlike [*sic*] computers would change the world," Rodney Brooks replied, in part:

> Since we won't have intelligent computers like humans for well over 100 years, we cannot make any sensible projections about how they will change the world, as we don't understand what the world will be like at all in 100 years.
>
> *(IEEE, 2017b)*

Although I do not always agree with Brooks, his point is an important one to remember in moving forward with the study of AI and its impact on the under-standing of the self. Brooks's comment echoes an ongoing critique of how we discuss and theorize AI as if it suddenly appears out of nowhere. This is another example of how AI is extraordinary. AI is unlike any other technology not just because it possesses humanlike elements but that by having these traits, AI is some-how magically exempt from the normal processes through which we make sense of our world. At least, that is how we have approached the study of AI and the self. For 70 years we have conceptualized AI as a technology that inherently challenges our very nature and our humanity. Yet, the AI that will allegedly lead to such onto-logical upheaval did not exist when this theoretical viewpoint was first introduced and still does not exist today. Our theory of what AI means for the self has been based on interpretations of contemporary technology projected onto the future.

But, as I have briefly demonstrated here, when we adopt an approach that situates the study of AI and the self within the context of our lived experiences, a different picture of AI appears. And this is what I mean by AI as ordinary. AI may be modeled on humans, thus, making it potentially different from other technolo-gies, but it is not exempt from the ways we make sense of our world. Further-more, no matter how humanlike technology becomes, our own sense of self also remains rooted in our interactions with the world around us. And so, regardless of the degree of the intelligence in the technology before us, to understand AI and ourselves, we must remember to go beyond the idea of AI as extraordinary, and approach it as ordinary.

Note

1 Most of the empirical work mentioned within this chapter is derived from my dis-sertation (Guzman, 2015), and some of the terms used here and discussions of findings may differ slightly from how I originally presented them within the dissertation. For example, I now use the term "vocal social agent," or VSA, to refer to the agents I stud-ied (see Guzman, 2017).

References

Agre, P. E. (1997). *Computation and human experience.* Cambridge: Cambridge University Press.

Anthes, G. (2017). Artificial intelligence poised to ride a new wave. *Communications of the ACM*, 60, 19–21. doi: 10.1145/3088342

Bloomfield, B. P., & Vurdubakis, T. (2008). IBM's chess players: On AI and its supplements. *The Information Society*, 24, 69–82. doi: 10.1080/01972240701883922

Blumer, H. (1969). *Symbolic interactionism: Perspective and method*. Englewood Cliffs, NJ: Prentice-Hall.

Boden, M. (2016). *AI: Its nature and future*. Oxford: Oxford University Press.

Bostrom, N. (2016). *Superintelligence: Paths, dangers, strategies*. Oxford: Oxford University Press.

Dechert, C. R. (Ed.). (1966). *The social impact of cybernetics*. New York: Simon & Schuster.

Dennett, D. C. (1988). When philosophers encounter artificial intelligence. *Daedalus*, 117 (Artificial Intelligence), 283–295.

Dreyfus, H. L. (1999). *What computers still can't do: A critique of artificial reason*, 6th ed. Cambridge, MA: MIT Press.

Dreyfus, H. L., & Dreyfus, S. E. (1988). Making a mind versus modelling the brain: Artificial intelligence back at the branchpoint. *Daedalus*, 117 (Artificial Intelligence), 15–34.

Ekbia, H. R. (2008). *Artificial dreams: The quest for non-biological intelligence*. New York: Cambridge University Press.

Ensmenger, N. (2012). Is chess the drosophila of artificial intelligence? A social history of an algorithm. *Social Studies of Science*, 42(1), 5–30. doi: 10.1177/0306312711424596

Franklin, S. (2014). History, motivations, and core themes. In K. Frankish & W. M. Ramsey (Eds.), *The Cambridge handbook of artificial intelligence*. Cambridge: Cambridge University Press.

Gandy, R. (1999). Humans versus mechanical intelligence. In P. Millican & A. Clark (Eds.), *Machines and thought: The legacy of Alan Turing*, vol. 1 (pp. 125–136). Oxford: Oxford University Press.

Goffman, E. (1959). *The presentation of self in everyday life*. New York: Anchor Books.

Graubard, S. R. (1988). Preface to the issue "Artificial Intelligence." *Daedalus*, 117 (Artificial Intelligence), V–VIII.

Guzman, A. L. (2015). *Imagining the voice in the machine: The ontology of digital social agents*. Chicago: University of Illinois at Chicago.

Guzman, A. L. (2016). The messages of mute machines: Human-machine communication with industrial technologies. *communication +1*, 5(1). http://scholarworks.umass.edu/cpo/vol5/iss1/4

Guzman, A. L. (2017). Making AI safe for humans: A conversation with Siri. In R. W. Gehl & M. Bakardjieva (Eds.), *Socialbots and their friends: Digital media and the automation of sociality* (pp. 69–82). New York: Routledge.

Guzman, A. L. (In Press). What is human-machine communication, anyway? In A. L. Guzman (Ed.), *Human-Machine Communication: Rethinking Communication, Technology, & Ourselves* New York: Peter Lang.

Haugeland, J. (1985). *Artificial intelligence: The very idea*. Cambridge, MA: MIT Press.

Holdren, J. P., & Smith, M. (2016). *Preparing for the future of artificial intelligence*. Washington, DC: National Science and Technology Council, Committee on Technology. www.datascienceassn.org/sites/default/files/Preparing%20for%20the%20Future%20of%20Artificial%20Intelligence%20-%20National%20Science%20and%20Technology%20Council%20-%20October%202016.pdf

IEEE. (2017a). Can we copy the brain? *IEEE Spectrum: Technology, Engineering, and Science News*, May 31. http://spectrum.ieee.org/static/special-report-can-we-copy-the-brain

IEEE. (2017b). Human-level AI is right around the corner—or hundreds of years away. *IEEE Spectrum: Technology, Engineering, and Science News*, May. http://spectrum.ieee. org/computing/software/humanlevel-ai-is-right-around-the-corner-or-hundreds-of-years-away

Kroeker, K. L. (2011). Weighing Watson's impact. *Communications of the ACM*, 54(7), 13. doi: 10.1145/1965724.1965730

Kurzweil, R. (2005). *The singularity is near*. New York: Penguin.

Leverhulme Centre for the Future of Intelligence. (2017). About: Preparing for the age of intelligent machines. http://lcfi.ac.uk/

McCarthy, J., Minsky, M. L., Rochester, N., & Shannon, C. E. (2006). A proposal for the Dartmouth summer research project on artificial intelligence, August 31, 1955. *AI Magazine*, 27(4), 12.

McCorduck, P., Minsky, M., Selfridge, O. G., & Simon, H. A. (1977). History of artificial intelligence. In *IJCAI* (pp. 951–954). http://ijcai.org/Proceedings/77–2/Papers/083. pdf

Mead, G. H. (1934). Obstacles and promises in the development of the society. In *Mind, Self and Society*. Chicago: University of Chicago Press, 317–327.

Moor, J. (2006). The Dartmouth College artificial intelligence conference: The next fifty years. *AI Magazine*, 27(4), 87.

Müller, V. C. (2016). New developments in the philosophy of AI. In V. C. Müller (Ed.), *Fundamental issues of artificial intelligence*, vol. 376 (pp. 1–4). Switzerland: Springer.

Neff, G., & Nagy, P. (2016). Automation, algorithms, and politics: Talking to bots: Symbiotic agency and the case of Tay. *International Journal of Communication*, 10, 17.

Neisser, U. (1966). Computers as tools and as metaphors. In C. R. Dechert (Ed.), *The social impact of cybernetics* (pp. 71–93). New York: Simon & Schuster.

Papacharissi, Z. (2011). *A networked self*. New York: Routledge.

Simon, H. A. (1999). Machine as mind. In P. Millican & A. Clark (Eds.), *Machines and thought: The legacy of Alan Turing*, vol. 1 (pp. 81–102). Oxford: Oxford University Press.

Theobald, R. (1966). Cybernetics and the problems of social reorganization. In C. R. Dechert (Ed.), *The social impact of cybernetics* (pp. 39–70). New York: Simon & Schuster.

Turing, A. M. (1948). Intelligent machinery. In D. C. Ince (Ed.), *Mechanical intelligence* (pp. 107–128). New York: North-Holland.

Turing, A. M. (1950). Computing machinery and intelligence. In D. C. Ince (Ed.), *Mechanical intelligence* (pp. 133–160). New York: North-Holland.

Turkle, S. (1984). *The second self*. New York: Simon & Schuster.

Winston, P. H. (1987). Artificial intelligence: A perspective. In W. E. L. Grimson & R. S. Patil (Eds.), *AI in the 1980s and beyond: An MIT survey* (pp. 1–12). Cambridge, MA: MIT Press.

8

AGENCY IN THE DIGITAL AGE: USING SYMBIOTIC AGENCY TO EXPLAIN HUMAN–TECHNOLOGY INTERACTION

Gina Neff and Peter Nagy

Recent advancements in technology challenge our fundamental notions of human power and agency. Tools and techniques including machine learning, artificial intelligence, and chatbots may be capable of exercising complex "agentic" behaviors (Dhar, 2016). Advanced technologies are capable of communicating with human beings in an increasingly sophisticated manner. Ranging from artificial chat partners through the commercial algorithms of social media to cutting-edge robots, these encounters with interactive machines often result in a complex and intimate relationship between users and technologies (Finn, 2017). For instance, people now may have their own virtual assistants such as Apple's Siri and Amazon's Alexa. Other commercial technological agents "help" people find new movies on Netflix or friends on Facebook. Robotic companions, like Huggable developed by Cynthia Lynn Breazeal at MIT, can read human emotions and react to them accordingly. Chatbots have long evoked reactions from people that can be used therapeutically for psychological counseling and now these tools are being rolled out as apps to help people cope with anxiety and depression (Lien, 2017; Neff & Nagy, 2016). Such developments present a quandary for scholars of communication. Does the agency of the people and, increasingly, of things that we choose to communicate with matter? This question prompts us to urge communication scholars to develop a better definition of agency and more clarity on how agency is enacted in practice within complex, technologically mediated interactions.

Agency has been inherently fuzzy, referring to different concepts depending on the particular discipline where it is being used. Campbell (2005) notes that "agency is polysemic and ambiguous, a term that can refer to invention, strategies, authorship, institutional power, identity, subjectivity, practices, and subject positions, among others" (p. 1). *Agency* has long connoted a distinctive human ability and something that by definition only humans possess. Theories connecting agency to intentionality (Jansen, 2016) and to the capacity to achieve one's goals (Caston, 2011) amplify this sense of agency. The idea of agency as inherently social and cooperative is constrained by the symbolic and material elements of context and culture (Campbell, 2005).

Recent social theories have started to analyze how non-humans, like technologies, may acquire agentic capacities and quasi-agentic abilities. For instance, when discussing the pervasive social implications of these emerging technological artifacts, communication and media studies scholars often focus on how and for what ends people use these tools. As such, they emphasize the role of social context and/or affordances in technology usage (e.g. Balsamo, 2011; Cooren, 2010; Siles and Boczkowski, 2012). Yet, communication and media scholars tend to neglect the different ways users imagine and communicate with these interactive social agents (Guzman, 2016; Leonardi & Barley, 2008). Science and technology studies (STS) scholars have already mapped how human and technological agents interact with each other. Bruno Latour (1999) from the Actor-Network Theory (ANT) movement, and François Cooren (2010) from the Montreal School of organizational communication have theorized the tensions between human and non-human actors. Similarly, communication and media scholars have focused on the different ways users derive meaning from encounters with technologies capable of exercising a wide range of agentic behaviors (e.g. Grossberg, 1997; Foss et al., 2007). What these conceptualizations still miss is that they have not addressed the ontological differences, and similarities, between human and non-human agency (Jansen, 2016).

As a corrective, we reframe and redefine agency in terms of how and when people interact with complex technological systems. For this we turn to Albert Bandura's social-cognitive psychological theory of human agency and to the concept of symbiosis in biology. Bandura's agency theory helps us understand the role of users in shaping and determining technical systems. Being an agent means that people can exert intentional influence over their mental processes and physical actions to achieve their desired outcomes (Bandura, 2006). People, however, rarely act on their own. They tend to delegate tasks to sophisticated technologies (e.g. smartphones, computers, etc.), often viewing them as personal assistants and companions rather than simple tools. Rod and Kera (2010) have specifically used the term "symbiosis" for our increasing interdependent relationships with technologies. The theory that we extend with this chapter imagines human–technology interaction as a symbiotic interplay of human and non-human agents, further blurring the line between what users want to do and what technologies are capable of doing. In the present chapter, we will introduce our symbiotic agency theory and show how it can address the problems around agency theories developed by communication and media studies as well as in science and technology studies.

Conceptualizing Agency in Human–Technology Interactions: An Unfinished Project

Agency is generally viewed as a capacity to act, produce, and anticipate a desired outcome within a particular context (Giddens, 1979; Wang, 2008; Leonardi, 2011). This contextual element to scholarly definitions mean agency is "communal, social, cooperative, and participatory and, simultaneously constituted and constrained by the material and symbolic elements of context and culture" (Campbell, 2005, p. 3). Communication theories tend to focus on exploring how users construe meaning through reflective and symbolic manipulation of technologies (Siles & Boczkowski, 2012). As such, communication and media scholars have made real progress in moving beyond the former simplistic views of human–technology interaction in which humans are always in control and technologies do not exert influence over their users (Orlikowski, 2005; Lievrouw, 2013; Guzman, 2016). However, communication and media studies theories often stop short of conceptualizing interdependent, entwining human, and non-human agencies, and instead focus on the agentic dimensions of socio-technical systems (e.g. Leonardi & Barley, 2008).

For STS scholars, agency is typically seen as people's general ability to shape the material form and meaning of artifacts through practices and interactions (Pickering, 1995; Feenberg, 1999; Gillespie, 2006). The STS literature discusses agency in conjunction with materiality or the content of media technologies, framing agency as a dynamic notion centered around how users create and interpret as well as shape and attach content to media technologies. As Latour (1999) phrased this, "we are sociotechnical animals and each human interaction is sociotechnical" (p. 214). By this he meant people delegate tasks to technologies, enroling them into larger scripts for action. Traffic speed bumps take the place of police officers standing at intersections, brought into action by a series of connected links to urge drivers to slow down. From Latour's perspective, agency can always be best understood as a complex interplay of non-humans and humans in these larger systems. Humans make non-humans act on their behalf, but non-humans also do the same with humans (Hannemyr, 2003).

The larger project of STS and ANT deepened our understanding of the social dimensions and social "lives" of technologies. Both communication and STS scholars have shown how agency is embedded in a broader sociocultural landscape. It also led, unfortunately, to what we might argue are simplified explanations of the role of technology as a social force, and a confusing conceptualization of agency. Bringing in language to address the multifaceted nature of human agency and its connections to technological agents would help correct this.

Our pragmatic working definition of agency is what users, actors, and tools, *do* when interacting with complex technological systems. This means that agency is closely linked to intentionality (Jansen, 2016) and to the capacity to achieve one's goals (Caston, 2011). Agency can mobilize people to reflect on their experience, adapt to their surroundings, and affect the sociocultural context that in turn sustains them (Jenkins, 2008). From this starting point, we extend the notion of agency by combining Albert Bandura's agency theory and the concept of symbiosis. This will help scholars develop a more nuanced and detailed understanding of how people think, feel, and act when interacting with the technical.

Defining Human Agency from a Social–Cognitive Perspective

Albert Bandura, one of the most influential social–cognitive thinkers, considers human agency a central force in human functioning (Bandura, 1989). Being an agent means that people can exert intentional influence over their mental processes and physical actions to achieve their desired outcomes (Bandura, 2000). Agency enables individuals to exercise greater control over their mental functioning and to monitor how their behaviors affect the world around them (Bandura, 2001). The social cognitive perspective holds that human agency is built around four core properties (Bandura, 2006). The first is intentionality, referring to action plans and potential strategies people form to achieve their goals and aspirations. The second property is forethought, plans humans fabricate for the future, helping them set goals and anticipate likely outcomes; thus forethought plays an important role in guiding and motivating people's efforts. In addition, agents are not only planners, but also self-regulators who are capable of managing their emotions and cognitions. Finally, the last component, self-reflectiveness, enables human beings to examine their own functioning through reflecting on their feelings, thoughts, and behaviors.

Since individuals rarely act on their own, social psychological theories also investigate the cognitive, motivational, or affective processes of agency not only on a personal but also on a social level. In addition to the previously discussed personal agency, people may exercise proxy and collective agency as well (Bandura, 2000). Whereas collective agency represents people's shared beliefs in accomplishing goals through cooperative efforts, proxy agency is enacted through different processes (Bandura, 2006). Proxy agency refers to a socially mediated mode that involves other individuals or tools that can act on their behalf, and can help someone to achieve goals and desirable outcomes. In this sense, proxy agency heavily involves the mediative efforts of others (Bandura, 2001). Finally, human beings mainly turn to proxy control to exert their influence when they lack developed means to do so or when they believe others can perform better (Bandura, 2002).

Proxy agency becomes a particularly powerful lens for looking at the relationships between users and tools. Consider the example of mobile phones. They require regular maintenance that necessitates human interventions such as charging the batteries, switching on/off, and paying the bills. But mobile phones also assist their users by reminding them about important upcoming events, providing information on issues, and ultimately helping them better organize their social lives. People can "outsource" their concerns about themselves to their devices, allowing their smartphones and other trackers to assist them in their efforts to get more sleep, more exercise, or fewer calories (Schüll, 2016). The concept of proxy agency helps to frame these desires and interactions as a form of power of remaking and reshaping the world that people and their tools attribute to one another.

By borrowing the idea of symbiosis, we illustrate the ways that fundamental biology concepts may help us conceptualize this entanglement of human and technological agency when we talk about complex sociotechnological systems.

The Concept of Symbiotic Agency

The term "symbiosis" derives from Greek meaning "living together." It was first coined by the botanist Anton de Bary in the mid-nineteenth century, who defined symbiosis as "the living together of different species" (Sparrow, 1978). As Sapp (1994) put it "we evolved from, and are comprised of, a merger of two or more different kinds of organisms living together. Symbiosis is at the root of our being" (p. xiii). As such, symbiosis is often viewed as a virtually ubiquitous phenomenon serving as the basis of normal development (Pradeu, 2011). Forming symbiosis may help organisms augment their robustness and functionalities (Kitano and Oda, 2006) and acquire new capacities (Douglas, 1994).

Psychologically informed research also viewed symbiosis as an essential concept and considered it a central force in orchestrating our relationship with the social world (Chodorow, 1989; Tissaw, 2000). Rather than being passive recipients, individuals are actively involved in their social development via taking part in symbiotic relationships. The idea of symbiosis can be applied to the human–technology relationship by illustrating the fact that people use technologies for augmenting their capacities or acquiring new ones. The previously discussed core properties of agency coined by Bandura (2006) can help us understand how symbiotic agency is formed between users and technologies.

In terms of *intentionality*, when individuals use their mobile phones as alarm clocks to wake them up in the morning, they rely on the perceived non-human agentic properties and affordances of mobiles. The user's intention to wake up in time is delegated to the mobile phone, in other words, they practice proxy agency. *Forethought*, on the other hand, gives us valuable insights into how users' future-oriented and anticipatory behaviors may be supported and extended by technological artifacts. For example, some individuals regularly use health-trackers or other related applications in hope of taking full control over their biological functioning. They do it because they would like to preserve their health, or prevent diseases. In other words, they use devices or applications as temporal extensions of their agencies. Again, similarly to intention, users delegate the forethought dimension of their agency in order to regulate their life, in this case their health, more effectively. They also use technological artifacts in the hope of creating coherence and finding meaning in their lives (Neff & Nafus, 2016), but perhaps more importantly, because health-trackers and related applications are interactive entities. The third agentic component, *self-regulation* (or as Bandura calls it self-reactiveness), points out the motivational and affective aspects of technology usage. For instance, several devices and programs are constructed to promote healthy behavior for users (e.g. quit smoking or healthy eating applications), for these artifacts send feedback on one's behavior and "performance" on a regular basis, mobilizing the motivational properties of agency. Human–technology interaction is characterized both by users' perceptions and emotions and by the emotions evoked by the design and aesthetic features of devices or digital products (Mick & Fournier, 1998; Thüring & Mahlke, 2007). Finally, with the emergence of sophisticated artificial agents, users gained ample opportunities to reflect on nearly every aspect of their lives (e.g. working, learning, or socializing). When applied to human–technology interaction, *self-reflectiveness* practices have been changed, shaped, and modified by technological artifacts due to the mediative characteristics of these tools. That is, users may explore technological artifacts socially, culturally, and cognitively so they can experience different modes of agentic functioning and self-reflective practices.

Tracing the forms of agency that people can have within technological systems, we propose the term *symbiotic agency*, referring to a specific form of proxy agency that users and tools can enact within human–technology interaction. We argue that *symbiotic agency* can be considered as a form of soft determinism encompassing both how technology mediates our experiences, perceptions, and behavior, and how human agency affects the uses of technological artifacts. More specifically, we propose that when individuals interact with technologies, they exercise proxy agency through a technologically mediated mode referring to the entanglement of human and non-human agencies. For symbiosis, similar to entanglement, implies an obligated relationship—symbionts are completely dependent on each other for survival, it can be considered a proxy agentic relationship that may provide different benefits for the recipients.

The Different Forms of Symbiotic Agency

Previous biological research also showed that symbiosis is a complex phenomenaon; organisms may benefit from, be harmed by, or not be affected by symbiotic relationships (Paracer and Ahmadjian, 2000). An association in which one symbiont benefits and the other is neither harmed nor benefited is a commensalistic symbiosis. An association in which both symbionts benefits is a mutualistic symbiosis. A relationship in which a symbiont receives nutrients at the expense of a host organism is a parasitic symbiosis. The conceptual boundaries, however, are not clear, and there are frequent transitions between the three forms of symbiosis (Tissaw, 2000). The different forms of symbiosis can be also used as a lens for understanding how users' expectations, beliefs, and values may influence their proxy agentic relationship with technologies.

People have a general tendency to anthropomorphize technologies that are often designed to elicit a wide range of projections (Darling, 2014). The "Uncanny Valley Effect" (Reichardt, 1978) documents the discomfort that people feel when robotic agents are seen as having a too humanlike appearance, and this effect is even more intense when a particular robot has some form of disfigurement or abnormal feature (Seyama & Nagayama, 2007). When ascribing agency to complex technologies, such as chatbots, robots, or other AI applications, people project their emotions to these entities and try to find explanations for their "behaviors" (Darling, 2014). A recent study by Neff and Nagy (2016) already showed that when users interact with chat programs or chatbots they tend to attribute "personality" to the technical, and project a wide range of emotions on to them, that in turn changes how the artifact communicates with them. Similarly, within media psychology, "computers as social actors" theory (Nass & Moon, 2000) holds that users unconsciously apply social rules to technologies and interact with them as if they were living organisms. The ways individuals approach these technologies can be traced further back to their general tendency to treat simple things as autonomous and intelligent entities (Shermer, 2008). For instance, in Fritz Heider and Marianne Simmel's (1944) classic experiment, participants were asked to watch a short animation about triangles and circles moving in weird patterns. Surprisingly, when instructing the subjects to describe what they saw, Heider and Simmel found that people fabricated complete and detailed stories about the geometric figures which they treated as agents and ascribed various emotional states and goals to them. These examples show why people may view their relationship with technologies as a form of commensalistic symbiosis and treat devices as simple tools enabling them to achieve their goals. Similarly, users may feel that they form mutualistic relationship with technologies, such as chatbots or robots, that simultaneously help them as well as learn from them. Finally, users may also feel that they are increasingly getting "addicted to" or "controlled by" technologies due to the perceived parasitic attributes of these tools.

Symbiotic agency and its different forms show us that human–technology relationship is no longer related solely to human action but also to technological action. Our concept represents the entanglement of human and technological agencies. We have argued here that symbiotic agency is a more appropriate expression to capture the dynamic nature of human and technological agency in human–machine communication, in which users simultaneously influence and are influenced by technological artifacts.

The Implications of Symbiotic Agency for Human–Technology Interaction

By building on our working definition of agency, we proposed a new theoretical framework for studying human–technology interaction. Our theory presents human–technology interaction as a symbiotic interplay of human and non-human agents, further blurring the line between what users want to do and what technologies are capable of doing. By combining psychological as well as communication and media studies concepts, we proposed that scholars can build a better theory for studying complex communication technologies. Technological artifacts have an impact on their users because of their interactive features and affordances (Nagy & Neff, 2015), and individuals' agentic properties shape the uses of artifacts for accomplishing goals and creating meaning. Reimagining agency this way brings meaning making into more arenas of social and technological life.

Symbiotic agency also helps scholars better capture the intimate linkages between human and technical agents. Ranging from artificial chat partners through the commercial algorithms of social media to cutting-edge robots, today's world is now filled with advanced technologies that are capable of communicating with human beings in a sophisticated manner (Guzman, 2016). When studying sociotechnical systems, symbiotic agency allows scholars to move beyond simplistic terms of the debate about social constructivism vs. technological determinism, and better conceptualize the encounters between human and non-human. With the notion of symbiotic agency, scholars are able to address and explore the dynamic dimensions of human–technology relationship. By redefining human–technology interaction as a unique form of proxy agency, media and communication theorists can investigate both the human agentic properties and the technological attributes of what it means to be a networked self.

Acknowledgments

This material is based upon work supported by the National Science Foundation under Grant No. 1516684.

References

Balsamo, A. (2011). *Gendering the technological imagination*. Durham, NC: Duke University Press.

Bandura, A. (1989). Human agency in social cognitive theory. *American Psychologist*, 44(9), 1175–1184.

Bandura, A. (2000). Exercise of human agency through collective efficacy. *Current Directions in Psychological Science*, 9(3), 75–78.

Bandura, A. (2001). Social cognitive theory: An agentic perspective. *Annual Review of Psychology*, 52, 1–26.

Bandura, A. (2002). Growing primacy of human agency in adaptation and change in the electronic era. *European Psychologist*, 7(1), 2–16.

Bandura, A. (2006). Toward a psychology of human agency. *Perspectives on Psychological Science*, 1(2), 164–180.

Campbell, K. K. (2005). Agency: Promiscuous and protean. *Communication and Critical/Cultural Studies*, 2(1), 1–19.

Caston, J. (2011). Agency as a psychoanalytic idea. *Journal of the American Psychoanalytic Association*, 59(5), 907–938.

Chodorow, N. J. (1989). *Feminism and Psychoanalytic Theory*. London: Yale University Press.

Cooren, F. (2010). *Action and agency in dialogue: Passion, incarnation and ventriloquism*. Philadelphia: John Benjamins.

Darling, K. (2014). Extending legal protection to social robots: The effects of anthropomorphism, empathy, and violent behavior towards robotic objects. In R. Calo, M. Froomkin, & I. Kerr (Eds.), *Robot law* (pp. 212–232). Camberley, UK: Edward Elgar.

Dhar, V. (2016). The future of artificial intelligence. *Big Data*, 4(1), 5–9.

Douglas, A. E. (1994). *Symbiotic interactions*. Oxford: Oxford University Press.

Feenberg, A. (1999). *Questioning technology*. New York: Routledge.

Finn, E. (2017). *What algorithms want: Imagination in the age of computing*. Cambridge, MA: MIT Press.

Foss, S. K., Waters, W. J. C., & Armada, B. J. (2007). Toward a theory of agentic orientation: Rhetoric and agency in Run Lola Run. *Communication Theory*, 17(3), 205–230.

Giddens, A. (1979). *Central problems in social theory: Action, structure, and contradiction in social analysis*. London: Macmillan.

Gillespie, T. (2006). Designed to "effectively frustrate": Copyright, technology and agency of users. *New Media & Society*, 8(4), 651–669.

Grossberg, L. (1997). *Bringing it all back home: Essays on cultural studies*. Durham, NC: Duke University Press.

Guzman, A. (2016). The messages of mute machines: Human–machine communication with industrial technologies. *Communication +1*, 5, Article 4. http://scholarworks.umass.edu/cgi/viewcontent.cgi?article=1054&context=cpo.

Hannemyr, G. (2003). The internet as hyperbole: A critical examination of adoption rates. *The Information Society*, 19(2), 111–121.

Heider, F., & Simmel, M. (1944). An experimental study of apparent behavior. *The American Journal of Psychology*, 57, 244–259.

Jansen, T. (2016). Who is talking? Some remarks on nonhuman agency in communication. *Communication Theory*, 26(3), 255–272.

Jenkins, A. H. (2008). Psychological agency: A necessarily human concept. In R. Frie (Ed.), *Psychological agency: Theory, practice, and culture* (pp. 177–200). Cambridge, MA: MIT Press.

Kitano, H., & Oda, K. (2006). Self-extending symbiosis: A mechanism for increasing robustness through evolution. *Biological Theory*, 1(1), 61–66.

Latour, B. (1999). *Pandora's hope: Essays on the reality of science studies*. Cambridge, MA: Harvard University Press.

Leonardi, P. M. (2011). When flexible routines meet flexible technologies: Affordance, constraint, and the imbrication of human and material agencies. *MIS Quarterly*, 35(1), 147–167.

Leonardi, P. M., & Barley, S. R. (2008). Materiality and change: Challenges to building better theory about technology and organizing. *Information and Organization*, 18(3), 159–176.

Lien, T. (2017). Depressed but can't see a therapist? This chatbot could help. www.latimes.com/business/technology/la-fi-tn-woebot-20170823-htmlstory.html?utm_campaign=KHN%3A%20First%20Edition&utm_source=hs_email&utm_medium=email&utm_content=55559892&_hsenc=p2ANqtz-8OaM-wjlJ3xD6NliosxXQXAwYWuYGCObgZbgysXOFjQlewbzhp8yJ3ueYtP2XlQu1N-ZDEFj7LyuTD6A9FBLTI_YcBCm0w&_hsmi=55559892

Lievrouw, L. (2013). Materiality and media in communication and technology studies: An unfinished project. In T. Gillespie, P. Boczkowski, & K. Foot (Eds.), *Media technologies: Essays on communication, materiality, and society* (pp. 21–52). Cambridge, MA: MIT Press.

Mick, D. G., & Fournier, S. (1998). Paradoxes of technology: Consumer cognizance, emotions, and coping strategies. *Journal of Consumer Research*, 25(2), 123–143.

Nagy, P., & Neff, G. (2015). Imagined affordances: Reconstructing a keyword for communication theory. *Social Media + Society*. http://sms.sagepub.com/content/1/2/2056305115603385.full

Nass, C., & Moon, Y. (2000). Machines and mindlessness: Social responses to computers. *Journal of Social Issues*, 56(1), 81–103.

Neff, G., & Nafus, D. (2016). *Self-tracking*. Cambridge, MA: MIT Press.

Neff, G., & Nagy, P. (2016). Talking to bots: Symbiotic agency and the case of Tay. *International Journal of Communication*, 10, 4915–4931.

Orlikowski, W. J. (2005). Material works: Exploring the situated entanglement of technological performativity and human agency. *Scandinavian Journal of Information Systems*, 17(1), 183–186.

Paracer, S., & Ahmadjian, V. (2000). *Symbiosis: An introduction to biological associations*. Oxford: Oxford University Press.

Pickering, A. (1995). *The mangle of practice: Time, agency, and science*. Chicago: University of Chicago Press.

Pradeu, T. (2011). A mixed self: The role of symbiosis in development. *Biological Theory*, 6, 80–88.

Reichardt, J. (1978). *Robots: Fact, fiction and prediction*. London: Thames & Hudson.

Rod, J., & Kera D. (2010). From agency and subjectivity to animism: Phenomenological and Science Technology Studies (STS) approach to design of large techno-social systems. *Digital Creativity*, 21(1), 70–76.

Sapp, J. (1994). *Evolution by association: A history of symbiosis*. Oxford: Oxford University Press.

Schüll, N. (2016). Data for life: Wearable technology and the design of self-care. *BioSocieties*, 11(3), 317–333. doi: 10.1057/biosoc.2015.47

Seyama, J., & Nagayama, R. S. (2007). The uncanny valley: Effect of realism on the impression of artificial human faces. *Presence*, 16(4), 337–351.

Shermer, M. (2008). Patternicity: Finding meaningful patterns in meaningless noise: Why the brain believes something is real when it is not. www.scientificamerican.com/article/patternicity-finding-meaningful-patterns/

Siles, I., & Boczkowski, P. (2012). At the intersection of content and materiality: A texto-material perspective on the use of media technologies. *Communication Theory*, 22(3), 227–249.

Sparrow, F. K. (1978). Professor Anton de Bary. *Mycologia*, 70, 222–252.

Thüring, M., & Mahlke, S. (2007). Usability, aesthetics and emotions in human–technology interaction. *International Journal of Psychology*, 42(4), 253–264.

Tissaw, M. A. (2000). Psychological symbiosis: Personalistic and constructionist considerations. *Theory & Psychology*, 10(6), 847–876.

Wang, Y. (2008). Agency: The internal split of structure. *Sociological Forum*, 23(3), 481–502.

9

THE IMMERSIVE VR SELF: PERFORMANCE, EMBODIMENT AND PRESENCE IN IMMERSIVE VIRTUAL REALITY ENVIRONMENTS

Raz Schwartz and William Steptoe

It was about 10:30 am when I got a message from Will saying: "What are you up to this lunch break? Let's go check out the new Oculus demo." An hour later I was inside a dark demo room donning a Rift headset and holding a pair of Touch controllers. A few seconds later I was in ToyBox, a social VR experience that facilitates 1:1 interactions in immersive VR. I saw an avatar in front of me waving hello and then I heard the voice coming from the avatar's mouth. The first few words were: "Hi Raz, how cool is this?!" It was Will. I looked down and saw a graphic representation of my hands that matched my physical hand movement. I moved my hands around and watched how accurately the avatar's hands moved. I waved hello back and gave Will a thumbs-up. "Yes, this is pretty neat." Will then reached out to a virtual ping-pong racket and handed it over to me. Soon enough we were playing ping-pong in VR. And although Will was in a completely different location, I felt like we were there together in that ToyBox room.

Both Will's and my avatars were rudimentary but even with these low-fidelity graphic bodies, our avatars facilitated a representation of self that was fundamentally different than other online avatars. The combination of high sense of body ownership with sense of presence (Slater, 2009) resulted in this inquiry into the unique characteristics that construct the "Immersive VR Self" and the contribution that it has to the sense of "being there."

With the development of technology and the rising spread of VR headsets, we see a growing offering of immersive VR applications that facilitate social interactions. Applications ranging from multi-player games (such as Star Trek: Bridge Crew and Lone Echo) to live broadcast events (*Bill Nye the Science Guy* on AltSpace and live sports on NextVR) and general chat rooms such as Oculus Rooms and Facebook Spaces utilize VR avatars as a foundational part of the experience. In each of these applications a user is either assigned with a platform avatar (an avatar they created beforehand and that can be used across applications) or are presented with an app-level avatar editor option that allows them to customize their appearance.

Avatars in immersive VR environments are different. Prior to immersive VR systems such as non-immersive virtual worlds, the control of avatars generally used indirect input devices such a mouse and keyboard and thus did not reflect an immersed user's body actions. With immersive VR systems, the body movements and signal they emit are immediately translated into a VR environment with the goal to mimic and create identical online patterns of physical actions. Consequently, therefore the participant's proprioceptive sense matches that of the virtual body movements.

The goal of this work is to provide a conceptual framing of this unique construct of an online self. We detail the differences between immersive and non-immersive VR environments, survey existing related academic work that touches upon the presentation of self in online networks as well as experiments that examined the body illusion that is facilitated by immersive VR systems. Finally, we coin the term "Immersive VR Self," define its special characteristics and its relation to the concept of presence.

Background

Social VR environments are not a new concept. Since the early 1980s and the pioneering but now primitive text-based MUD (Multi-user Dungeons) systems, researchers have studied and developed technologies to support remote social interaction in online virtual worlds (Ito, 1997; Ren et al., 2007; Slater & Sanchez-Vives, 2016). The rise of the internet and 3D graphics technology contributed to the popularization of rich online virtual communities attracting millions of users such as Second Life and World of Warcraft. In Second Life people used their desktop computer to log in to a networked platform hosting a vast user-created virtual world, complete with its own currency and tools for evolving that world. In this environment, users are able to create an avatar based on their likings and to start a digital life there while meeting people, buying and selling goods, and exploring new places.

Until recently, this non-immersive mode of interfacing with virtual worlds has been the norm. Over the last few years, however, immersive VR systems that bring the user into the virtual environment with a wide field-of-view stereoscopic and perspective-correct rendering have started to gather traction among the general public. Companies such as Facebook, Oculus, HTC, Sony, and Google are investing resources in both hardware and software development and launching consumer products that aim to make immersive VR experiences more approachable and affordable.

Over the years, tech professionals, researchers, and fans of the field heavily promoted the gospel of "Being There" that is facilitated by this technology (Slater et al., 1994; Barfield et al., 1995; Rheingold, 1991). Immersive VR systems require the user to wear a tracked headset that constructs the sense of presence. Just like a magic trick, when a person puts on the headset, both their visual and audio senses are manipulated by this technology. This manipulation can create an experience that might teleport them to a different location such as a different country or provide stimuli to make them feel like they are in a different situation with other people or creatures (e. g. having a T-Rex running toward you while you are visiting the Natural History Museum in London).

Immersive vs. Non-Immersive Virtual Reality Environments

How are immersive VR systems, such as the Oculus Rift, different from non-immersive desktop-based VR? The fundamental difference is in the way this technology tracks a user's body movement and how its displays drive our perceptual senses. Immersive VR presents perspective-correct, wide field-of-view and stereoscopic imagery, which, together with spatialized audio, more closely mimics the natural and egocentric visual and aural sensory stimuli through which humans perceive the physical world. It does this by tracking the user's body movement, including hand movement, in real time. Hence, natural sensorimotor actions can be performed to explore, move around, and experience the virtual environment. This in particular is what gives rise to the feeling of "being there"—since if you perceive the world much as you do the real world, then the simplest perceptual hypothesis for the brain is "this is where I am" (even though you know it is not true). It is an illusion, but like other illusions knowing that it is an illusion does not make it go away.

Non-immersive VR systems do not directly track body motion, and so must rely on indirect interfaces, such as keyboard and mouse, to capture user input. Additionally, the displays do not surround the user and so they can only present a narrow viewport into the virtual environment.

From the visual aspect, to navigate in a non-immersive virtual environment (VE) such as Second Life, the user sits in front of a standard computer screen and uses a mouse or keyboard to look around. In addition to the narrow viewport, the common view in Second Life and other virtual worlds is not from the eyes of the user but rather from the avatar's proximity that includes a bigger area of the environment (although some virtual worlds allow a first-person view, these are generally not the default). As stated above, an immersive VR system provides rich three-dimensional imagery, head-tracked wide field-of-view, and controllers that mimic hand gestures.

In immersive VR, audio feedback is spatialized based on the user's head position in relation to the virtual audio source. For example, going back to the T-Rex scenario, when the dinosaur is far away, the sound of its roar and breathing is much quieter than when it looms over me. Spatialized audio also exists in non-immersive VEs but the combination of spatialized audio, head tracking, and immersive visual display is what makes the audio stimuli feel convincing and real.

Finally, head and hand gestures in immersive VR are driven directly from tracked body movement. Unlike non-immersive VR, the real-time tracked movement of users is mapped directly onto their virtual avatar. In this way, as a user nods their head or waves, their avatar replicates the same movement with a direct mapping.

In summary, the combination of head tracking to drive the egocentric, stereoscopic, and wide field-of-view into the VE, gesture-based input, and spatialized audio presents an experience that differentiates immersive VR from non-immersive. Immersive VR can result in the perceptual sensation of *presence*, which is best defined as a user's psychological response to patterns of sensory stimuli, resulting in the user having the illusion of "being there," in a computer-generated space (Slater et al., 1994).

Studying the Self in Immersive Virtual Reality

Existing research into the presentation of self in immersive VR has mostly looked at the technological- and interaction-based questions involving the design and behavior of avatars. However, it is hard to think about self-presentation in immersive VR without understating the broader landscape of scientific inquiry into the presentation of the self-online.

Today it is almost a cliché to cite Goffman's pivotal work on understanding the presentation of self in every daily life (Goffman, 1959). Unsurprisingly, Goffman's concepts of frontstage and backstage during the performance of self in social interactions as serving the need to control impressions can also be applied to immersive VR where people engage in online interactions using their avatars (Schroeder, 2002).

Academic investigation into the presentation of self and the role of avatars in online virtual works has been going on for decades now. From early works that looked at the construction and reconstruction of identity and the options to have multiple identities (Turkle, 1994), to studies into gender bending and how and why people customize their avatar in multi-player online games (Ducheneaut et al., 2009; Yee et al., 2011).

Experiencing the self in a virtual environment can happen either by having technology that recreates our actual bodies from the physical world or by artificially constructing them (Biocca, 1997; Mantovani, 1995). In either way, no matter if a user has set their avatar to look like their physical self or an imagined, completely different, alter-self, the self-identification with their VR representation plays a major part in the feeling of an existence of a virtual self and embodiment (Lee, 2004).

As defined previously, embodiment relates to a combination of sensations that appears in conjunction with being in, having, and controlling a virtual body (Kilteni et al., 2012). The academic investigation into this idea has been going on for several decades now and is primarily trying to understand what are the conditions that construct sense of embodiment and what are the different ways these conditions come together.

More specifically, studying embodiment in social interactions in VR, researchers suggested the idea of Transformed Social Interaction (TSI), which defines three dimensions that can differ entiate online representation of the self from a physical representation: self-representation, sensory abilities, and situational context (Bailenson et al., 2008). For example, researchers had participants use different colored bodies (black and white) to test issues such as racial bias (Banakou et al., 2016). In this work, an experiment was designed to examine if the perceptual illusion of body ownership reduced implicit bias. The results of this research show that implicit bias decreased after going through the experiment and particularly only for white participants in the black body.

In another project, researchers conducted experiments that showed the existence of a "Proteus Effect," which describes how a person's behavior aligns with their avatar design (Yee and Bailenson, 2007). This research found that when participants were assigned with attractive and tall avatars they behaved in a different way. More specifically, the research ran two experiments, the first study assigned attractive avatars to participants and as a result showed people engaged in more intimate conversations as well as closer distance than people who used less attractive avatars. In the second study, people who were given a taller avatar were more confident in negotiations than people with shorter avatars. A follow-up study looked at the priming effects of avatars and found that the look of the avatars primed their thoughts and ideas (Peña et al., 2009).

Another seminal work in the field of body ownership in immersive VR reproduced the rubber hand illusion and showed that people's perception of their virtual body as their physical body is feasible (Slater et al., 2008, 2009). Using a series of experiments that included projecting a virtual arm and using a data glove and a brain computer interface (BCI), the results suggested that virtual limbs and bodies in virtual reality can have a strong sense of ownership.

The tail end of this research recap describes a project about virtual tails on avatars (Steptoe et al., 2013). In this experiment participants had an avatar that was enhanced by a tail that moved around. One group of participants had a tail that moved randomly and the other group could potentially learn how to control the tail using their hip movement. The researchers found that people that were in the group that could control the tail had a higher level of body ownership and responded to threats to their virtual tail more strongly than those whose tail moved randomly.

Discussion

As we can see, the above-mentioned research in immersive VR focused on how people perceive and view their own avatar as well as the derivative effects of the avatars on their own physical behavior. This past research exemplifies how many concepts from the physical world translate directly to immersive VR environments. Ideas around performance, sense of self, embodiment play out in a similar manner and result in perceptions and actions that construct the notion of presence.

In this chapter, however, we would like to look at a more nuanced part of what supports the idea of presence. This is the construct of the self in immersive VR systems and what are its unique characteristics. As Slater clearly articulates, presence is constructed by Place Illusion (PI) and Plausibility Illusion (PSI) (Slater, 2009). Place Illusion creates the sensation of being in a real place while plausibility illusion creates the belief that the situation people are experiencing is actually occurring. More specifically, Slater sees the body as "a focal point where PI and PSI are fused." By which he means, looking down and seeing the graphic representation of your hands or body makes you feel like you are there (PI). But the fact that this virtual body also moves and responds in a certain logical way supports the plausibility illusion.

At this point, we would like to consider not only the actual presence and movement of the body as part of the factors that contribute to presence but also the sense of self that is created in that situation. Just like previous characteristics of online presentation of self, the immersive VR self carries the same traits but also incorporates three additional factors that distinguish- itself from previous online presentation of self:

- *Visual*: perspective-correct visual representation.
- *Audio*: spatialized sounds.
- *Movement*: physical body gestures.

These three elements contribute to the enhanced sense of self that is enacted in immersive virtual environments. It can be described as shown in Figure 9.1.

It is important to note that Figure 9.1 describes the perception of avatars in immersive VR from the spectator point of view. This means, these factors come into play when we see another avatar in front of us (or our own avatar in a mirror). When these three dimensions work as whole in an immersive VR environment, they construct what we perceive as a representation of a person. In other words, Figure 9.1 depicts the observable signals that construct the Immersive VR Self, each part is connected and affected by the others to create the presentation of oneself.

Visual. First, the visual depiction of the self. The fact that immersive VR provides a 3D and spatial view of the environment operates both on the perception of my body and on other people's presence. In several immersive VR systems, when a user looks down, they see a representation of their own body and hands. When looking at other avatars, their body occupies a volumetric space just like in the physical world. This visual representation of my personal body as well as others is the first part of the triangle. This part is affected by the two other parts: the

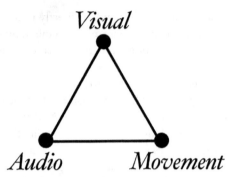

FIGURE 9.1   The three distinguished dimensions of the Immersive VR Self

body moves based on the physical gestures of people and in some cases the mouth animation based on the user's talk.

Audio. The audio is another crucial part of the presentation of self in immersive VR. The voice channel that feeds into the avatars is a live audio stream of the person's live talk. In this way, the person's voice is enacted by the avatar. This provides users with the option to hear the voice of the person behind the avatar and their non-verbal utterances. This part is affected by the visual presentation as the audio is spatialized—if the visual and movement presentation is of an avatar that is far away or avatar that looks to the right or left, the audio will be heard correspondingly to be far or coming from one of the sides accordingly.

Movement. The final part of Figure 9.1 includes movement and gestures; movement data is being translated from the person's physical actions. For example, if a person is waving hello with their hand or pointing toward a direction with their finger, their avatar creates a similar action in the VR environment.

Sense of self is achieved when these three factors work in tandem. The Immersive VR Self therefore portrays these unique characteristics that combine visual, audio, and movement signals from the physical world to construct an embodied representation in an online environment. A level of self-presentation can still be achieved with only one or two of these but the more signals the avatar can perform, the higher is the sense of presence and of self.

As we show in this chapter, physical observable signals that are in turn translated into graphic signals in the VR environment create a unique representation of the self that is different from other online examples. Avatars in immersive VR environments are mimicking physical signals in a fidelity that did not exist before. Unlike previous venues of online social interaction, it is harder to fake the behavior of the avatar in immersive VR. In many cases the movement of the head as well as other body gestures react naturally to various stimuli and these reactions are reflected instantly online.

If we go back to the definition of presence (Slater, 2009), we argue that the Immersive VR Self is part of the illusion that supports the sense of being there.

Together with the body, place, and plausibility of the scenario, the person's performance and perception of the sense of self constructs the experience of the immersive VR environment. Until now, the majority of past academic research efforts gave little attention to theory of self-presentation and how it manifests itself in immersive VR. Our chapter suggests that, when studying "Presence," in addition to looking at the technical perspectives of place, actions, and body, we must take into consideration the performance and perception of the self.

Conclusion

In this chapter, we examine the presentation of self in immersive VR environments. First, we describe the difference between an immersive and non-immersive system and then survey academic research that studies different perspectives of embodiment and sense of presence in immersive VR. Finally, we propose the notion of the "Immersive VR Self" and identify the three factors that make it a unique construct of online self-presentation.

Future work in this field should extend the investigation of this presentation of self. For example, the role of the fluidity of the settings in which we experience the avatars is an interesting topic to understand. In this way, being the same avatar in different contexts might replicate similar notions of context collapse we already saw appear in online social networks (Hogan, 2010).

As we are in early stages of mainstream VR systems, in the next few years we will see a growing adoption and use of these headsets, and as a result, increasing venues in which the Immersive VR self will perform. Studying and understanding the sociological forces that drive embodiment and sense of self in VR can help researchers further examine behavioral and group interaction dynamics as well as inform developers and practitioners designing immersive VR environments that play to the strengths of this form of embodiment.

References

Bailenson, J. N., Yee, N., Blascovich, J., & Guadagno, R. E. (2008). Transformed social interaction in mediated interpersonal communication. *Mediated Interpersonal Communication*, 6, 77–99.

Banakou, D., Hanumanthu, P. D., & Slater, M. (2016). Virtual embodiment of white people in a black virtual body leads to a sustained reduction in their implicit racial bias. *Frontiers in Human Neuroscience*, 10, 601. doi: 10.3389/fnhum.2016.00601

Barfield, W., Zelter, D., Sheridan, T., & Slater, M. (1995). Presence and performance within virtual environments. In W. Barfield & T. A. Furness (Eds.), *Virtual environments and advanced interface design* (pp. 473–513). Oxford: Oxford University Press.

Biocca, F. (1997). The cyborg's dilemma: Progressive embodiment in virtual environments. *Journal of Computer-Mediated Communication*, 3(2), 1–31.

Ducheneaut, N., Wen, M.-H., Yee, N., & Wadley, G. (2009). Body and mind: A study of avatar personalization in three virtual worlds. *Proceedings of the 27th international conference on human factors in computer systems* (pp. 1151–1160). New York: ACM.

Goffman, E. (1959). *The presentation of self in everyday life*, Anchor Books edition. Garden City, NY : Doubleday. https://search.library.wisc.edu/catalog/999467804702121

Hogan, B. (2010). The presentation of self in the age of social media: Distinguishing performances and exhibitions online. *Bulletin of Science, Technology & Society*, 30(6), 377–386.

Ito, M. (1997). Virtually embodied: The reality of fantasy in a multi-user dungeon. In D. Porter (Ed.), *Internet culture* (pp. 87–109). Routledge.

Kilteni, K., Groten, R., & Slater, M. (2012) The sense of embodiment in virtual reality. *Presence: Teleoperators and Virtual Environments*, 21, 373–387.

Lee, K. M. (2004). Presence, explicated. *Communication Theory*, 14(1), 27–50.

Mantovani, G. (1995). Virtual reality as a communication environment: Consensual hallucination, fiction, and possible selves. *Human Relations*, 48(6), 669–683.

Peña, J., Hancock, J.T., & Merola, N.A. (2009). The priming effects of avatars in virtual settings. *Communication Research*, 36(6), 838–856. https://doi.org/10.1177/0093650209346802.

Ren, Y., Kraut, R., & Kiesler, S. (2007). Applying common identity and bond theory to design of online communities. *Organization Studies*, 28(3), 377–408.

Rheingold, H. (1991). *Virtual reality: Exploring the brave new technologies*. New York: Simon & Schuster.

Schroeder, R. (2002). Social interaction in virtual environments: key issues, common themes, and a framework for research. In *The social life of avatars: Presence and interaction in shared virtual environments* (pp. 1–18). London: Springer.

Slater, M. (2009). Place illusion and plausibility can lead to realistic behaviour in immersive virtual environments. *Philosophical Transactions of the Royal Society B: Biological Sciences*, 364(1535), 3549–3557. doi: 10.1098/rstb.2009.0138

Slater, M., & Sanchez-Vives, M.V. (2016). Enhancing our lives with immersive virtual reality. *Frontiers in Robotics and AI*, 3, 74.

Slater, M., Usoh, M., & Steed, A. (1994). Depth of presence in virtual environments. *Presence: Teleoperators & Virtual Environments*, 3(2), 130–144.

Slater, M., Perez-Marcos, D., Ehrsson, H. H., & Sanchez-Vives, M. V. (2008). Towards a digital body: The virtual arm illusion. *Frontiers in Human Neuroscience*, 2, 6. doi: 10.3389/neuro.09.006.2008

Slater, M., Perez-Marcos, D., Ehrsson, H. H., & Sanchez-Vives, M.V. (2009). Inducing illusory ownership of a virtual body. *Frontiers in Neuroscience*, 3(2), 214–220. doi: 10.3389/neuro.01.029.2009

Steptoe, W., Steed, A., & Slater, M. (2013). Human tails: Ownership and control of extended humanoid avatars. *IEEE Transactions on Visualization and Computer Graphics*, 19(4): 583–590.

Turkle, S. (1994). Constructions and reconstructions of self in virtual reality: Playing in the MUDs. *Mind, Culture, and Activity*, 1(3): 158–167.

Yee, N., & Bailenson, J. (2007). The Proteus Effect: The effect of transformed self-representation on behavior. *Human Communication Research*, 33(3), 271–290.

Yee, N., Ducheneaut, N., Yao, M., & Nelson, L. (2011). Do men heal more when in drag?: Conflicting identity cues between user and avatar. *Proceedings of the 2011 SIGCHI conference on human factors in computing systems* (pp. 773–776). New York: ACM.

10

WRITING THE BODY OF THE PAPER: THREE NEW MATERIALIST METHODS FOR EXAMINING THE SOCIALLY MEDIATED BODY

Katie Warfield and Courtney Demone

This is going to be a tough discussion so you might want to sit down. May I hold your hand? Okay…

It's not you. It's your research methods.

This chapter came about after three years of working in various capacities and with various projects studying how young people experience sharing images of their bodies on social media. Over the past three years I have conducted phenomenological interviews with young women taking selfies, online surveys with young men and women who share images on Facebook, Instagram, and Snapchat, and interviews and focus groups with trans and gender non-conforming people who share images depicting their gendered becoming via networked images on Tumblr. And there is a constant in all these projects. The constant is that image-makers and the images they produce are deeply and affectively connected. And it is really beautiful to listen to. One of my participants in the Tumblr project stated: "Some people reblog [selfies] if they think you look great or just want to support their friends. It's a way to spread positivity and support. It feels good." Another participant said, "I post selfies on Tumblr a lot … because it helps me feel connected to other users as a person, and it's an ego boost if I get a lot of notes, and I feel like I can be more authentic … with mostly strangers rather than people I know IRL. It's also just part of the culture of the website." Courtney, the co-author of this chapter, in writing about sharing her YouTube coming out video online said that she was "sharing a piece of [herself]." I asked her to elaborate and her response was, "It was absolutely a story that was a part of myself. This was me sharing a piece of myself with people close to me, with the intention of bringing a change to our relationships." The YouTube video was a visual representation, it was also a presentation of herself but it was moreover, and importantly, connected to her body and well-being. The socially mediated images of bodies young people share online are really networked, and not just technologically, but affectively, discursively, and materially.

In this chapter, what we want to propose is that the dominant research methods used to study socially mediated images of bodies often do not treat socially mediated images of bodies as closely connected to the bodies of the image-makers. It is important to note that we are not saying *researchers* are not treating bodies as affectively networked—wonderful research abounds that recognizes this[1]—but rather what we want to argue is that *the research methods themselves* discourage the idea that the image-makers and the images they make are fundamentally entangled. We argue that the prevailing research methods place and position the researcher on the *exterior of the image* and so at a distance from the image-takers. The distance created by the research method *itself* is shaping both how we think about socially mediated images of bodies and how we encourage other researchers to handle images of bodies. We draw on feminist new materialist theories to talk about *research assemblages* (Ringrose & Renold, 2014; Fox & Alldred, 2014) or the arrangement of objects in the empirical study and how *research methods themselves shape how the phenomenon presents itself, and subsequently how we think about the objects of inquiry*, or here: *mediated bodies*. We are going to argue that research methods that distance the researcher from the bodies of the image-makers also *distance our responsibility to those bodies*. In response, we are going to suggest that methods that bring researchers *closer to the image-maker and the images shared online* might provide both a more *proximate* insight into the lived experiences of the image-takers *and* a more responsible and ethical research conduct. We argue that the closer we bring researchers to the *affective materiality* or the *matter* of the image-takers' bodies and how those bodies are entangled with the images produced, the more these bodies may come to *matter* to the researcher.

And so our chapter will play out in the following manner: we are going to trace the dominant paradigms that underscore the prevailing methods used to study socially mediated bodies to show how the paradigms of *representationalism* (Barad, 2007) and *presentationalism* create distance and to a degree disregard between image-maker, image, and researcher. We then discuss how new materialism, predominantly via the work of Karen Barad, can provide an ethico-onto-epistemology (Barad, 2007, p. 185) that can replace *distance and disregard* with *proximity and respect*. We present and illustrate three new materialist-informed methods we have developed: 1) *meating* in the story; 2) networked intra-viewing; and 3) reading the image horizontally to bring proximity between image-maker, image, and researcher. We play with the words *meet* and *meat* to introduce a novel method that involves a co-authoring process that places an emphasis on listening to the embodied, affective, and bodily felt experiences of sharing images of the body online. *Networked intra-viewing* and *reading networked images horizontally* are both methods adapted for online environments from existing methods: intra-viewing from Kuntz and Presnell (2012) and reading images horizontally from

Hultman and Taguchi (2009). Courtney and I elaborate each of these three meth-ods with reference to our work together and to the various images Courtney has shared of herself online via Facebook, Instagram, and YouTube.

Literature Review

We would like to begin by proposing that two predominant paradigms underscore research methods used to study mediated images of bodies: *representationalism* and *presentationalism*. We borrow the concept of *representationalism* from Karen Barad, who says that the paradigm sees images as determinate objects that are separate and distinct from reality (Barad, 2007). Representationalism underscored the dis-cursive treatment of early studies of digital images, treating them as determinate and distinct objects akin to analogue photos of the Kodak brownie era (Munir & Phillips, 2005; Gye, 2009), early portraiture like the family photo (Haldrup & Larsen, 2003), and travel photography (Urry, 2007). Early analyses of photoblog-ging (Cohen, 2005) and analyses of early camera phone uses adopted methods of analysis that emerged from visual culture and semiotic traditions. These methods placed the researcher on the outside of the image and focused their attention on the surface of the image to indicate uses, subject matter, visual trends, and mean-ings in digital-imaging practices (Kindberg et al., 2005; Lister, 1995; van House, 2011; Frosh, 2015; Rettberg, 2014).

Presentationalism is a term we are using to mark research methods or frame-works that suggest that the self shared online via photo or video is a theatrical performance, a calculated presentation, and thus different from and often less *authentic* than the offline self. Research methods or frameworks that form this paradigm often make reference to the works of Judith Butler or Erving Goff-man, and presentational frameworks have been used to examine: the unequal gender policing that happens online to shape mediated images of bodies (Mar-wick, 2013), impression management of beauty bloggers on Instagram (Abidin, 2016), the empathy-inducing profile images of subaltern small business owners on micro-lending platforms (Gajjala, 2015), the miniaturizing corporeal poses of girls in selfies (Warfield, 2015), *self-pornification* poses in male nude selfies (Lasén, 2014).

We will now visualize the research assemblage of both these paradigms *to look at the arrangement between image-maker, images of the mediated body, and researcher.* If the socially mediated image of the body is *representational*, then the research method places the researcher on the outside of the image. The researcher hovers over it as if it is an object on a table. The photo-taker is often not considered a part of this research assemblage. The photograph or video, as object, is treated as solid, distinct, and separate from the person who produced it. Representationalism is researcher-centric as it is the researcher's senses that are privileged in the process of analysis and meaning making. Representational methods place the researcher in close proximity to the image but at a distance from the image-taker. Drawing on Karen Barad (2007), representationalism "does not disrupt the geometry that holds object and subject *at a distance*" (p. 88), and here it holds the image-maker at arm's length from the researcher.

Next we will look at the research assemblage for the *presentational* paradigm. If the mediated body is approached from the *presentational* paradigm, then the method and framework encourage researchers to toggle their position: researchers move from a position in front of the image, reading it for tropes and performative patterns, and then often move behind it, looking at the calculated production of the image by the image-taker. But layered in this is that presentational methods and frameworks presume the image is a *performance*. When researchers are in front of the image, they see a stage with a thespian, and when they step behind the image they examine the calculated, premeditated, scripted, production. Via Goffman, a researcher who adopts presentationalism sees there is no *real self*, only theatrical performances interactionally negotiated amidst sociocultural fields and forces. But what this paradigm also supposes is that such presentations are outside the realm of the *real* and, as such, are of lesser significance or regard. Even if a social media researcher is *theoretically* challenging aspects of the paradigms of representationalism and presentationalism by adopting inadvertently tried and tested research methods used in studying socially mediated images of bodies, the methods position them in predictable positions and relationships which inadvertently reinforce representationalism and presentationalism in the results.

So representationalism creates *distance* and presentationalism, to a degree, *disregard*. Are there methods that could encourage *proximity and respect*? I draw on new materialism to fill the proximity gap.

The material turn has many names and forms—posthumanism, new materialism, the post-qualitative turn—and have several core theoretical similarities but proximity and complex interconnections are a core component. These common themes include: the rejection of the Cartesian divide between mind and body (Barad, 2007), a critique of the linguistic turn's emphasis on language at the expense of materiality (Barad, 2007), a rejection of positivism's tendency to simplify and generalize via positivist linear cause and effect relations in attempts to organize the natural complexity of the world around us (Deleuze et al., 1984), and a critique of the researcher's presumed-to-be removed role from their objects of study and the research event itself (Barad, 2007).

Given its penchant for complexity, new materialist methods often begin with *complex units of analysis* like for Deleuze and Guattari the *assemblage*, and for Karen Barad the *phenomenon*. Instead of choosing *just* the photo or *just* the image-maker/performer, research methods often begin with the multifaceted entanglement of material and discursive forces in a given moment. This may include an inseparable coming together of bodies, subjectivities and audiences, and non-human actors like objects, technologies, space, and place (Latour, 1993). Metaphors of *assemblage* and *entanglement* are preferred over metaphors that differentiate and delineate one entity within a phenomenon from the other.

Why is unit of analysis in new materialist methods not distinct but rather complex? New materialists see objects not as a priori but rather as foundationally entangled and then *emerging* from their entanglements through processes of naming and differentiating (Deleuze et al., 1984). It is through what Barad (2007) calls *intra-actions—becoming through* not *interacting among*—that things like concepts and objects (like the *body*, *identity*, and the *image*) are defined and constantly redefined. Barad introduces the concept of the *agential cut* to describe the process by which objects and concepts are *cut from* their natural state, which she proposes is one of material and discursive entanglement and complexity (Barad, 2007).

Here we can illustrate the notion of the agential cut with our arguments about how representational and presentational research methods shape our understanding of socially mediated images of bodies. Because digital images materially *look like* analogue photos (Munir & Phillips, 2005; Gye, 2009; Haldrup & Larsen, 2003; Urry, 2011), they have become discursively aligned with fine art self-portraits (Frosh, 2015). The research methods that were adopted at the start of such analyses therefore *read* digital images the same way we would *read* analogue photos in the past: by assessing the surface of the image. The problem is that this neglects the fact that digital images are often produced from specific audiences on specific social media platforms, at specific moments in time to reflect specific messages and moods and are further shaped by the limitation in technological and platform affordances. Only a handful of these forces influenced the production of photos in the analogue era. When we *discursively treat* socially networked photos like analogue photos, we are performing a process of agential cutting wherein the discursive cookie cutter of analogue photo research methods is used on digital photos thereby privileging habitual research methods while in the process also snipping some of the natural complexity of digital images along the same lines as have been used to treat analogue photos. Agential cuts slice the complex nature of phenomena along habitual and normative lines, which in turn reinforces their discursive and material treatment as similar to objects and concepts analogous to them in the past.[2]

Agential cuts also affect the relationship between the researcher, image, and image-maker. Representation and presentational paradigms perform agential cuts separating the image from the producer, and orienting the gaze of the researcher either toward the image (representationalism) or influencing how serious or less sserious the researcher will interpret digital imaging of bodies online to be. In contrast to these paradigms, when we see research methods through a new materialist paradigm, there is no such thing as an a priori researcher and research participant cleaved from one another, but rather an already and always complex entanglement of material and discursive forces that comes together in any research assemblage. According to new materialism *there is no a priori space between researcher*, image, and image-maker. Everything is already touching. It is the researchers themselves, via age-old methods and habitual research practices of distancing who repeatedly *cut away and cut apart that proximity, closeness and responsibility* to create space between themselves, the images, and the image-maker. This cutting is what results in a process of *distancing* and *discounting* by researchers themselves of image-takers from their entangled lived experiences online.

Three New Materialist Methods

We would like to propose that the manufactured distance between image-maker, image, and researcher created by the representational and presentational paradigms may be contracted through three new materialist-informed methodologies that Courtney and I played with in our time together: 1) *Meating* in the story; 2) networked intra-viewing; and 3) reading networked images horizontally. *Meating* in the story brings proximity between researcher and image-maker, networked intra-viewing brings proximity between image and image-maker, and reading images horizontally rethinks images not only as flat texts but also as material discursive entanglements. We will discuss each of these methods with reference to Courtney and my work together.

Meating in the Story

To understand the method of *meating* in the story we need to know a bit more about Courtney and the different images of her body she had shared via social media platforms. Courtney Demone is a trans activist who used a series of online platforms (YouTube, Instagram, Mashable, etc.) to come out and for advocating the rights of trans people online. She shared her coming out video on YouTube in 2015. The video received positive support and negative transphobic comments. The video was shared non-consensually on 8chan, a popular MRA spin-off of 4chan hostile to feminists primarily but also often queer and trans folks as well as people of color. In the same year, Courtney launched a political project on Instagram in which she posted images of the slow physical development of her breasts while on hormone therapy. Her project, *Do I have boobs now?*, asked when Instagram would begin censoring her images, and demanded attention to the coding of the politics of gender into regulatory processes on social media. In essence, when the social media platform finally decided she passed, her body would then be censored and gender policed from public display.

To discuss the method of meating in the story, I focus on Courtney's coming out video posted on YouTube and how we adapted existing narrative methods for social media to bring proximity between image, image-maker, and researcher. In order to write Courtney's narrative, we wanted to move beyond the outside-in research position of representational methods via semiotic analysis, and presentational frameworks via social interactionalism, that have become prominent in narrative methods for social media (Georgakopoulou, 2016). Courtney and I set up a shared Google document in which she wrote an initial accompanying *backstory* to explain the visual story she posted on YouTube (Figure 10.1). Once the first

When I initially came out as a woman, I told my close friends and family face-to-face, but rather than let word spread through acquaintances and the story possibly deviate, I chose to post a video on YouTube and share it with my Facebook friends. Posting that video was nerve racking. I had come out to a handful of people so far, some who handled it really well, some who handled it extremely poorly. I had so much internalized shame and guilt about being trans and posting that video to Facebook made me feel really vulnerable, sharing a piece of myself that I hid my whole life out of fear, and that's only been received with mixed reactions so far. I was literally shaking when making it and posting it, and for the hours that followed. My partner at the time took me out of the house with me presenting femme for the first time, in hope that it would make me less anxious about what was happening on Facebook. It sorta backfired and I felt even more nervous, especially since I was now in public and not passing very well. When I finally sneaked a peek at my phone, there were dozens of lovely and supportive comments from my friends, even from those who I thought were going to take it poorly. I sorta felt like I had just done a trust fall: I had put myself in a very vulnerable place but my friends supported me anyway. I was still nervous and shaky, but I was also overwhelmed with love and kindness.

Within a day, I had a ton of wonderful comments on Facebook, but I received some emails notifying me of comments on my YouTube video. This was a little surprising, because I had kept the video as unlisted so only the people I shared the link with would be able to see it. When I checked the YouTube comments, there was a dozen or so horribly mean comments from people I didn't know saying I was delusional, that I should be ashamed of myself, that no one would ever see me as a woman, etc. Near the end of these comments, one of these strangers explained that a friend of mine (from the way it was worded, it was implied that this was a close

katie warfield
13 16 16 Jan   Resolve ⋮

Above you use the term "sharing my video" and here you use the term "sharing a piece of myself", I want to pick up those terms: How did you conceive of your youtube video? Was it a "story"? Was it disconnected from you? Was it a story and a part of you? I'm sort of trying to get at what was the emotional or embodied connection you felt when this "video" was shared without your consent and especially on platforms known for anti-feminist activities.
Show less

Courtney Demone
18:59 18 Jan

It was absolutely a story that was a part of myself. This was me sharing a piece of myself with people close to me, with the intention of bringing a change to our relationships. It felt very much like an unwelcome intrusion to have these other people listening and commenting.

FIGURE 10.1   Excerpt from YouTube story on sharing the body.

draft was complete, I met her in the shared document in real time and asked her to elaborate on sections of the plot, characters, places, audiences, and so on. After her story was developed, I then wrote a draft of a paper and posted it in a shared googledoc for Courtney. We then reversed the roles and Courtney read my article and commented on it, asking questions about my writing, interpretations, and uses of her stories and experiences. This process served both as a form of member checking and as a means to bring researcher closer to image and image-maker. I then altered the paper to reflect and respect her comments and questions.

Further we noticed that the affordances of the shared googledoc within the research assemblage were more similar to the medium of writing online than were we to do an offline interview. Also, the real-time aspect of this mode of data collection assured that we were able to encourage proximity between the story, Courtney's lived experience, and my role as a co-author and researcher. Thus what the platform affordances offered was what we could call "live member checking" *within* the actual story itself and this encouraged a closeness between storyteller, story, and researcher (me).

I want to note that this method of data collection, although novel, is not different from some other methods involving co-authoring and co-writing, particularly in narrative inquiry. The method we elaborated adopts a combination of already established methods: narrative co-authoring (Saltzburg and Davis, 2010) alongside real-time intra-viewing (Kuntz and Presnell, 2012). The method adds to narrative inquiry methods that have been developed by theorists like Alexandra Georgakopoulou but it pushed them beyond their representational and presentational paradigms which are present in her recent work on YouTube and selfies on Facebook (Georgakopoulou, 2016). This method shares commonalities with feminist narrative inquiry methods, which encourage participants to be "active agents" in order to bring proximity between researcher and storyteller (Fraser and McDougall, 2017, p. 242).

Networked Intra-Viewing

Although I was attentive to the different and interwoven plots of Courtney's stories, what stood out most was the presence of emotion, embodiment, and affect as central to her storytelling. Instead of weaving a story separate from and at a distance from herself, the story played out as very connected to her bodily experience and entangled within material and discursive forces. This reinforced the importance of working against the distance created between image and image-maker by representational and presentational paradigms. As a result, I drew on the new materialist method of *intra-viewing*, which places a particular emphasis on affect and proximity between participant and researcher. Whereas the narrative tradition often falls prey to the "narrative seduction" (Brinkmann, 2011) of the words and a focus almost exclusively on the humans in the research encounter (anthropocentrism), the intra-view attends to the human and non-human factors of the intra-view as well as the material and discursive forces entangled within and beyond the story.

Kuntz and Presnell (2012) list several tactics to intra-viewing: *a move from script to sound* where the researcher attends to not only the spoken words but also the pauses, tones, inflections, glitches, and mistakes in conversation, as well as the *embodied vibrations* where and when the body withdraws from conversation, or the text becomes distant and less intimate or equally proximate and increasingly intimate. I adapt this method for *networked intra-views* by recognizing that in messaging and typed online chatting, kinesics are often not visible and so verbal pauses often take the form of typing pauses where one person in the interaction dwells on the other person's words before responding. The pause does not necessarily mean disagreement but is an in-road for future questioning (e.g. a researcher might ask, "Did I notice a pause there? Did you have other thoughts?"). I also note that the intra-view technique of *embodied vibrations* in spoken conversation may also be more detectible as the researcher would be able to hear cues like changes in tone of voice. For online settings, intra-viewers can be attentive to similar qualities: changes in writing style, curt or abrupt sentences or the use of emoji for expressing different emotions.

Intra-views also attend to the use of *metaphors* to describe experience which Neisser (2003) argues are creative spaces. Metaphors offer a place for participants to break from linguistic norms and take language on an embodied line of flight (Deleuze et al., 1984) through words in attempts to narrate the momentary layering of the world on the skin. Metaphors often are attempts at describing the affective encounters of the participant within his/her/their worlds. In networked intra-views, online writing may already encourage more metaphorical description as the person is typing instead of talking. Whereas a person might censor the desire to use poetic language in spoken word, in writing, the modality permits more privacy and comfort to play with that form of communication. I used all these tactics in the intra-viewing process while meating Courtney in the story via the shared googledoc.

Networked Intra-Viewing

In this next section, Courtney elaborates on the non-consensual sharing of her coming out story on 8chan.

Following the new materialist methodologies proposed by Neisser (2003), I focus on the metaphor of becoming "worn down" that is first used to describe the experience of being harassed online after her video was shared to hostile social media platforms (Figure 10.2). When I ask her to elaborate on those metaphors, she provide an expansively detailed further metaphorical narrative and explanation that is very bodily in nature: she says her body was worn down, her strength, she describes the experience like water torture and slow drops on her forehead. The emotional impact of having the images of her body shared online was deeply physical, psychological, and emotional. This sort of assessment could not be achieved if the researcher held a position from the outside of the image. It is only when the researcher becomes close to the writing and the writer that this sort of data is revealed.

FIGURE 10.2 Excerpt of metaphors.

We also observed several moments of attending to embodied vibrations during the intra-view process conducted in the googledoc. I used this as an opportunity to ask more. I also used it occasionally as an opportunity to give her time to think. On one occasion I interpreted some of her words in the document and she paused and then wrote "Umm…" I prompted her by questioning my own interpretation as a researcher "did I misunderstand? Am I completely misinterpreting this?" She said yes and then explained what she meant by her words. This humbling dialogue brought Courtney and me closer together, corrected my misinterpretations of her writing, and provided more space for her voice.

Reading Networked Images Horizontally

Given Courtney's story is so entangled with visual technologies (video camera and camera) and technologies of networked sharing (YouTube, Instagram), it is important that research methods attend to agency of the materiality of research assemblage and not just the agency of words. It is not only the discourses that flow through technologies to affect outcomes, but also the design and presence of the technologies themselves that can have impacts on the research assemblage and outcomes. Within a new materialist framework then, agency is not singular and linear—discourse shapes action—rather agency is entangled and multiple originating from many sources and moving in manifold directions within the research assemblage.

For our study, and within a new materialist framework then, the image itself, how photos are typically handled, and the action potential of photos as research objects within the research assemblage, can wield agentic power to shape research outcomes. Karin Hultman and Hillevi Lenz Taguchi (2009) have elaborated what they call a *relational materialist* approach to reading visual data like photos or videos. They argue that instead of reading the object of the photo *anthropocentrically,* as photos habitually are, where we focus predominantly on the *humans* in the image, the process of reading the image horizontally marks human and non-human elements of the photo as equally important and agentic in the production of the image. In their original explanation, Hultman and Taguchi discuss a photo they captured and analyzed in a research project that examined childhood play. In the photo, a young girl sits in a sandbox and watches as sand flows between her fingers. Whereas an anthropocentric visual analysis would attend to the girl exclusively and her agency on the sand, a relational materialist analysis argues: "the girl and the sand have no agency of their own" rather agency is seen as:

> "emerging" in-between different bodies involved in mutual engagements and relations: muscles lifting the arm and hand which slowly opens up and lets go of the sand, which by the force of gravity falls with specific speed into the bucket. The uneven foundation of the sandbox forces the body

of the girl to adjust her whole body around the sand. The force of gravity, the uneven foundation, the bucket and the quality of the grains of sand are all active forces that intra-act with her body and mind and that she had to work with and against.

(Hultman & Taguchi, 2009, p. 530)

When we read photos horizontally, we move both beyond the humans in the photo to add *dimensionality and depth* to the universe that is operating in the production of the actions in the image.

We can use a relational materialist approach to look at Courtney's images she shared on Facebook which were subsequently censored. Whereas Hultman and Taguchi's child intra-acts agentically with the sand and the space of the sandbox, in one of Courtney's censored photos she takes the arrangement of a selfie: her attention is directed playfully toward the camera. I suggest that we can adapt the horizontal reading of photos to include the intra-action of Courtney's body and presentation of self with several material forces of the image in this moment. She intra-acts with while being affected by: the technology of the camera, the social media platform of networked sharing and the specific affordances of those platforms, the discursive regulatory mechanisms that control her gender presentation, and the material technological mechanisms that control the visibility and sharing of her images. Again she does not simply adopt gendered tropes in the presentation of the self, but she both is shaped, shapes, and works among and against many forces in this moment of photo production. Thus reading the image horizontally entangles the images beyond the borders of the photo among the materiality necessary in the production and dissemination of the image. The image then is not simply a representation or a presentation but also a technologically entangled material and discursive process of *becoming*.

Since networked photos are never static and often flow from one location online to another or are interacted with in a communicative way, I suggest that *reading networked images horizontally* also needs to attend to the temporal future and past of the lifecycle of the photo within its networked existence. To visualize this we can picture positioning ourselves as researchers not outside the image, but rather positioning ourselves at the edge of the image, our eye sitting halfway above and halfway below the plane of the image surface. We can picture the specific material and discursive elements of the photo reaching forward in time and backwards in time beyond the surface of the image: Courtney's body stretches into a past and a becoming future, the image itself existing in its original moment of posting but then censored at a later date. Reading a networked photo horizontally recognizes that an image has a temporal history and a temporal future that is entangled within the material discursive assemblage of social networks, technologies, audiences, gender norms, and platform affordances.

We can now position ourselves at the edge of Courtney's photo to look at the temporal past and history of its lifespan online in order to see the material and discursive regulatory forces that policed the existence of her photo and becoming online. Courtney tells the story:

> Only a few months after the project was launched, Facebook and Instagram chose to censor all my photos, even those where there was no breast development. I think this was spawned by a Guardian piece about my project where the journalist actually contacted Facebook and asked for their opinion. Shortly after that, all my photos were censored.

Here we see that the lifespan of the image connected to its networked distribution via mainstream media outlets (*The Guardian*). The publicity of the images in mainstream media brought visibility to her body, while informing audiences of her politics, this led to audiences flagging her posts, and the subsequent censoring of her photo. Entangled with the image is the online news outlet that initially distributed her photos, the mainstream media outlets that broadcast it to wider audiences, and the affordances that permitted these wide audiences to flag its appropriateness thus enforcing the gender policing of her body's visibility.

Subsequent to its widespread dissemination online, Courtney details how the photos' censoring is entangled among material affordances and discursive forces that detailed the "community standards" resulting in the gender policing of her body online. In Figures 10.3–10.10, Courtney provides visuals of the censored image, the censoring process, and the regulatory affordances that controlled the visibility of her images.

FIGURE 10.3 Making a post.

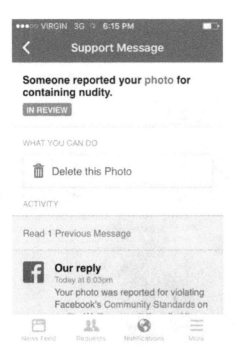

FIGURE 10.4 Getting reported by someone.

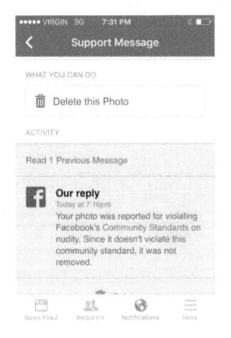

FIGURE 10.5 Algorithm gives it the OK.

We Removed Something Your Page Posted
We removed the post below because it doesn't follow the Facebook Community Standards:
Continue

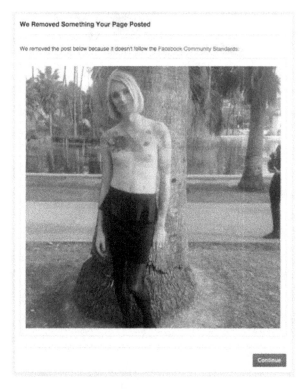

FIGURE 10.6 Reported again, and flagged by a moderator.

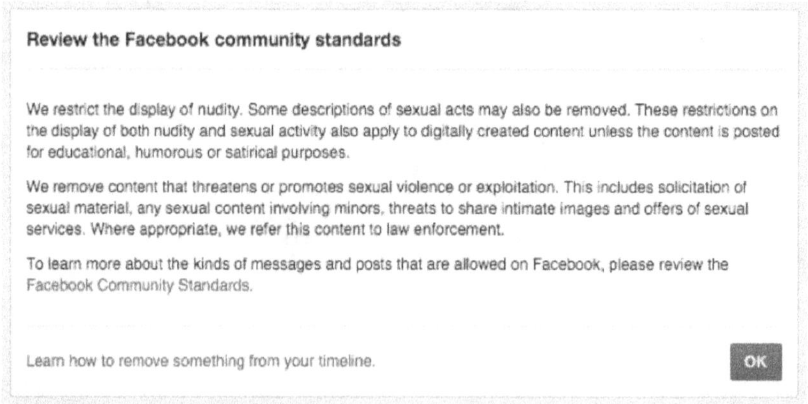

Review the Facebook community standards
We restrict the display of nudity. Some descriptions of sexual acts may also be removed. These restrictions on the display of both nudity and sexual activity also apply to digitally created content unless the content is posted for educational, humorous or satirical purposes.
We remove content that threatens or promotes sexual violence or exploitation. This includes solicitation of sexual material, any sexual content involving minors, threats to share intimate images and offers of sexual services. Where appropriate, we refer this content to law enforcement.
To learn more about the kinds of messages and posts that are allowed on Facebook, please review the Facebook Community Standards.
Learn how to remove something from your timeline.
OK

FIGURE 10.7 Review of Facebook community standards.

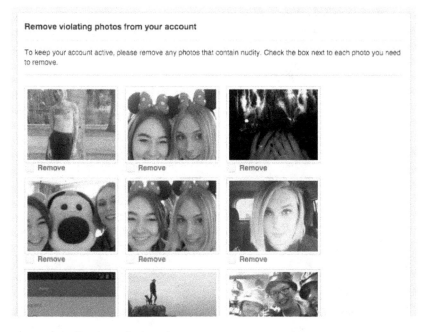

Remove violating photos from your account

To keep your account active, please remove any photos that contain nudity. Check the box next to each photo you need to remove.

Remove   Remove   Remove

Remove   Remove   Remove

FIGURE 10.8   Review of other photos.

You're Temporarily Blocked From Posting

This temporary block will last 3 days, and you won't be able to post on Facebook until it's finished.

Please keep in mind that people who repeatedly post things that aren't allowed on Facebook may have their accounts permanently disabled.

Continue

FIGURE 10.9   Temporary blocking.

We can see in this section of the photo's lifespan the entanglement of the material affordances of the regulatory handling of the photo with the discursive forces that police the visibility of Courtney's body in the image. A human reports the image deeming Courtney's body to be read as female, subject to gendered norm of visibility, and therefore inappropriately *nude*. A machine gives the image the okay, thus denying her being read as female, while also liberating her body from gendered norms of policing. A human moderator flags the photo again marking her body as being read as female and subjecting her body again to

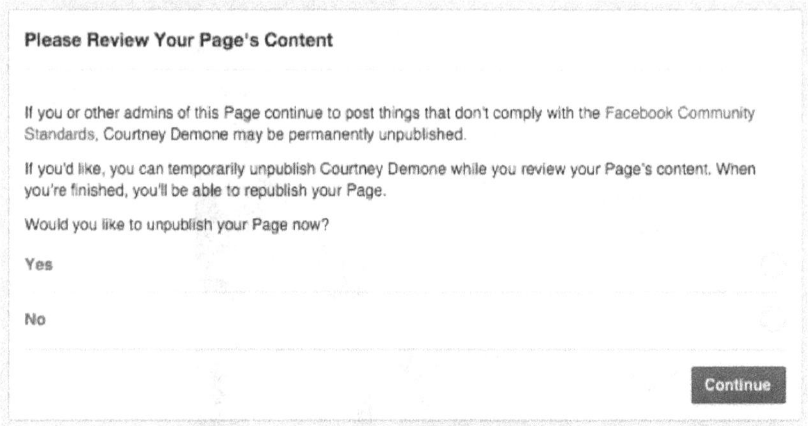

Please Review Your Page's Content

If you or other admins of this Page continue to post things that don't comply with the Facebook Community Standards, Courtney Demone may be permanently unpublished.

If you'd like, you can temporarily unpublish Courtney Demone while you review your Page's content. When you're finished, you'll be able to republish your Page.

Would you like to unpublish your Page now?

Yes

No

Continue

FIGURE 10.10   Requirement to review other photos on account.

gendered norms of visibility. A formulaic message is delivered to Courtney giving her veiled polite but limited options for action apart from removing the photos. The affordances of the flagging process provide Courtney no option for dispute, thus showing the rigid upholding of gendered regulatory mechanisms within the mechanics of the platform. The platform further requests that Courtney review her other photos suggesting that since the "standards" Courtney used to assess her original photo did not match the "community standards" of the platform, that the same "offending standards" may have been used to assess other images on her feed, thus forcing Courtney to embody the normative and regulatory community standards in a process of self-policing.

In a further conversation with Courtney, she expands on the tacitly gendered and trans-exclusive nature of Facebook's community standards:

> What did irk me was that both Instagram's and Facebook's messages they sent me when censoring my photos said that they were doing it to ensure a safe space on their platforms. Meanwhile, I was reporting all sorts of comments that were misgendering me, using slurs, and calling for violence against me, and all I was getting was "We reviewed this comment and found it doesn't violate our community guidelines." How do you define "safe" when nipples aren't safe and threats are?

The political sharing of images of women's breasts are censored upholding gendered norms of the visibility of women's bodies. Trans-exclusive comments are not censored as they are deemed by the platform as free speech and thus within the platform's community standards. When we look at the lifespan of networked images through a horizontal reading of networked images, we reveal many more material and discursive forces at play than simply what is seen at the surface.

Conclusion

Whereas presentational and representational paradigms create distance between researcher, image, and image-taker, I draw on new materialist methods, and adapt them for online researchers, to encourage a *contracting of the space* between writer, image, and researcher. Courtney and I did this by proposing three new materialist-influenced methods: 1) *meating* in the story; 2) networked intra-viewing; and 3) reading networked images horizontally.

Meating in the story is a play on meeting in the story but rather than focusing exclusively on the discursive data, the process emphasizes the material and discursive qualities of the writing itself. We wanted to move beyond the outside-in research position of representational and presentational methods and so adopted a method that placed us in close proximity in the process of storytelling and meaning making. As I mentioned, I wanted to contract the space between researcher, writer, and writing and I wanted to play with methods that respected the mediated *becoming* of Courtney's story. Meating in the story involves adopting a writing method that mimicked the material affordances of the original dialogic writing environment in which Courtney first posted her stories: YouTube and Instagram. It also invited Courtney to be present in both the data collection and the writing up of the chapter itself.

Networked intra-viewing draws on offline intra-viewing methods, which encourages attentiveness to not just the verbal information but also the *metaphors* and *embodied vibrations* of the participant such as pauses, omissions, and changes in tone. We adapt these methods for online intra-viewing and suggest that listening can occur between the words in live-interviewing practices by noticing pauses in responses and changes in writing style that may reflect changes in mood. Networked intra-viewing also attends to the use of metaphors which we argue may be even easier to encourage since the affordances and modality of writing online provide the safety and privacy that might encourage that type of expression.

Reading networked images horizontally is a method adapted from Karin Hultman and Hillevi Lenz Taguchi's (2009) model of reading analogue photos horizontally. Developed from a new materialist framework that privileges human and non-human elements in a photo, reading networked photos horizontally attends to the way agency works in a dispersed, entangled, and multi-directional way in a photo. When it comes to networked images, this method attends to the technology used to take the image (cameraphone, video, etc.), the platform used to distribute the image (YouTube, Facebook, etc.), and how those forces are entangled with the intersectional discursive forces in play which may involve gender, race, class, sexuality, ability, and so on. The method of reading images horizontally also attends to the lifespan of the photo, which may reveal that the importance and impact of these different material and discursive forces may apply more or less force at different moments in time in the lifespan of the photo.

All three methods challenge representational and presentational paradigms by starting from the point that research phenomena are not discrete moments to be studied by an objective researcher at a distance but rather are complex material and discursive *becomings* in which the researcher is always and already entangled. As such, and in this case, Courtney's images of her body must be studied as a series of complex assemblages entangled with identity, material bodies, emotions, affect, technologies of networked sharing, and audience practices of dialogue and response.

Importantly, my point is not to suggest that the paradigms of representationalism and presentationalism do not have validity. Representationalism provides us with in-roads to examine the mechanisms of power that work to enforce hegemonic conceptions of socially mediated bodies and presentationalism brings attention to the negation of selves amidst the push and pull of forces associated with identity categories. These paradigms have important merit. What I am suggesting, instead, is that the ontological strength of these paradigms is so rooted, and the arrangement of researcher to research subjects the paradigms encourage so embedded, that these paradigms are shaping the studying and handling of socially mediated bodies by creating distance between data and data producer. I argue that this could have ethical implications on how we as researchers come to handle online data, particularly data connected to the body, and that new materialist-informed methods could encourage a more ethical handling of data at the level of the individual researcher and more broadly at the level of research ethics.

Notes

1 For research that recognizes images as connected affectively to the image producer see: the networked self (Papacharissi, 2011) and the way time and space are felt online (Georgakopoulou 2017), as well as ample literature discussing the affective connection between young people and images they share online (Tiidenberg, 2014; Hillis et al., 2015; Vivienne, 2016, 2017; Raun 2012; Warfield 2016).
2 For a more extended discussion of what forces and how material discursive forces create agential cuts, see Barad, 2007.

References

Abidin, C. (2016). "Aren't these just young, rich women doing vain things online?" Influencer selfies as subversive frivolity. *Social Media & Society*, 1(4), 1–17.

Barad, K. (2007). *Meeting the universe halfway: Quantum physics and the entanglement of matter and meaning*, 2nd ed. Durham: Duke University Press.

Brinkmann, S. (2011). Interviewing and the production of the conversational self. In N. Denzin & M. Giardina (Eds.), *Qualitative inquiry and global crises* (pp. 56–75). Walnut Creek, CA: Left Coast Press.

Cohen, K. R. (2005). What does the photoblog want? *Media, Culture & Society*, 27(6), 883–901.

Deleuze, G., Guattari, F., & Massumi, B. (1984). *A thousand plateaus: Capitalism and schizophrenia*, 9th ed. Minneapolis: University of Minnesota Press.

Fox, N. J., & Alldred, P. (2014). New materialist social inquiry: Designs, methods and the research-assemblage. *International Journal of Social Research Methodology*, 18(4), 399–414.

Fraser, H., & McDougall, C. (2017). Doing narrative feminist research: Intersections and challenges. *Qualitative Social Work*, 16(2), 240–254.

Frosh, P. (2015). The gestural image: The selfie, photography theory, and kinesthetic sociability. *International Journal of Communications*, 9, 1607–1628.

Gajjala, R. (2015). Crafting the Digital Subaltern 2.0 through Philanthropy 2.0: Global Connection and access through digital micro lending. Presented at the Pixelated Eyeballs Lecture Series, Kwantlen Polytechnic University, March. www.youtube.com/watch?v=23lNl3JkBV0

Georgakopoulou, A. (2016). From narrating the self to posting self(ies): A small stories approach to selfies. *Open Linguistics*, 2(1), 300–317.

Georgakopoulou, A. (2017). Small stories research: A narrative paradigm for the analysis of social media. In L. Sloan & A. Quan-Hasse (Eds.), *SAGE handbook of social media research methods* (pp. 266–282). London: Sage.

Gye, L. (2009). Picture this: The impact of mobile camera phones on personal photographic practices. *Continuum*, 21(2), 279–288.

Haldrup, M., & Larsen, J. (2003). The family gaze. *Tourist Studies*, 3(1), 23–46.

Hillis, K., Paasonen, S., & Petit, M. (2015). *Networked affect*. Cambridge, MA and London: MIT Press.

Hultman, K., & Taguchi, L. (2009). Challenging anthropocentric analysis of visual data: A relational materialist methodological approach to educational research. *International Journal of Qualitative Studies in Education*, 23(5), 525–542.

Kindberg, T., Spasojevic, M., Fleck, R., & Sellen, A. (2005). The ubiquitous camera: An in-depth study of camera phone use. *IEEE Pervasive Computing*, 4(2), 42–50.

Kuntz, A., & Presnell, M. (2012) Wandering the tactical: From interview to intraview. *Qualitative Inquiry*, 18(9), 732–744.

Lasén, A. (2014). "But I haven't got a body to show": Self-pornificaiton and male mixed feelings in digitally mediated seduction practices. *Sexualities*, 18(5–6), 714–730.

Latour, B. (1993). *We have never been modern*. Cambridge, MA: Harvard University Press.

Lister, M. (Ed.). (1995) *The photographic image in digital culture*. New York: Routledge.

Marwick, A. (2013) "Gender, sexuality and social media." In Senft, T. & Hunsinger, J. (eds), *The social media handbook* (pp. 59-75). New York: Routledge.

Munir, K. A., & Philips, N. (2005). The birth of the "Kodak moment": Institutional entrepreneurship and the adoption of new technologies. *Organization Studies*, 26(11), pp. 1665–1687.

Neisser, J. (2003). The swaying form: Imagination, metaphor, embodiment. *Phenomenology and the Cognitive Sciences*, 2, 27–53.

Papacharissi, Z. (2011). *A networked self*. New York: Routledge.

Raun, T. (2012) Out online: Trans self-representation and community building on YouTube. PhD dissertation, Roskilde University.

Rettberg, J. W. (2014). *Seeing ourselves through technology: How we use Selfies, blogs and wearable devices to see and shape ourselves*. London: Palgrave Pivot.

g segments.

Ringrose, J., & Renold, E. (2014). "F**k rape!": Exploring affective intensities in a feminist research assemblage. *Qualitative Inquiry*, 20(6), 772–780.

Saltzburg, S., & Davis, T. (2010). Co-authoring gender-queer youth identities: Discursive tellings and retellings. *Journal of Ethnic and Cultural Diversity in Social Work*, 19(2), 87–108.

Tiidenberg, K. (2014). Bringing sexy back: Reclaiming the body aesthetic via self-shooting. *Cyberpsychology: Journal of Psychosocial Research on Cyberspace*, 8(1).

Urry, J. (2007). *Mobilities*. London: Sage.

van House, N. A. (2011). Feminist HCI meets Facebook: Performativity and social networking sites. *Interacting with Computers*, 23(5), 422–429.

Vivienne, S. (2016). *Digital identity and everyday activism: Sharing private stories with networked publics*. Basingstoke: Palgrave MacMillan.

Vivienne, S. (2017). "I will not hate myself because you cannot accept me": Problematizing empowerment and gender diverse selfies. *Popular Communication*, 15(2), 126–140.

Warfield, K. (2015). The model, the #realme, and the self-conscious thespian. Digital subjectivities in the selfie. *International Journal on the Image*, 6(2), 1–16.

Warfield, K. (2016). Making the cut: An agential realist examination of selfies and touch. *Social Media + Society*, 2(2), 1066–1079.

11

CLONES AND CYBORGS: METAPHORS OF ARTIFICIAL INTELLIGENCE

Jessa Lingel

Part of the anxiety around artificial intelligence (AI) stems from its liminality, and the contingent inability to disambiguate human from non-human, to identify the artificiality of sociotechnical intelligence. As a genre that alternately predicts and ridicules, science fiction offers many tropes and models for making sense of tensions surrounding human relationships to technology, including those around AI. Drawing on science fiction as well as STS theory, I compare two metaphors for their sensemaking capacity of AI: cyborgs and clones. As AI continues to evolve, it will become increasingly necessary to develop narratives that allow us to make sense of the divergences and convergences between human and non-human. Science fiction, and particularly (I will argue) patterns of twins and clones in science fiction, can help navigate shifting relationships between people and technologies as AI continues to integrate into everyday life.

I start from the premise that science fiction offers a productive alternative to other more rationalist, obtuse, and beholden-to-generally-accepted-standards-of-reality narratives of relationships between people and technology. Anecdotally, I have noticed an increased attention among academics and activists toward science fiction. Book clubs that used to read only academic texts are beginning to incorporate speculative fiction. Artists and technologists are convening science-fiction conferences and even live-action role-playing games informed by dystopian science-fiction plots. While one might attribute this interest to a desire for escapism or in response to a surge in high-quality science fiction in mainstream entertainment media, my interpretation is different: that people who think about ethical relationships to technology are grasping for a vocabulary and set of narratives that can help make sense of rapid sociotechnical change. Why should we be limited to strict and narrow concepts of reality (i.e. what currently exists mass market) in discourses of technology, given that change is rapid and continual? Why emphasize the rational and proven over the speculative and imagined? Assuming we are not trying to explain precisely how a technology functions but only how people relate to it, why not turn to science fiction as a genre with a telos of predicting, unpacking, and manipulating social perceptions of scientific and technological change?

Nakamura (2002) coined the term *cybertype* "to describe the distinctive ways that the Internet propagates, disseminates, and commodifies images of race and racism" (p. 3). Sinnreich and Brooks (2016) offered an extension of Nakamura's thoughts to produce the term *futuretype*, meant to

> encompass a broader range of issues of difference, and new political actors and subjects, with a focus on the ways in which ideological expectations, assumptions, and biases are encoded into the stories we tell ourselves and one another about the future of our species and life in the universe.
>
> *(p. 5664)*

Their project (on which I collaborated) involved looking to science fiction to help make sense of continually shifting and socially complex technological landscapes. Contributors considered why certain tropes, like alien encounters (Lichfield et al., 2016) or immortality (Adams et al., 2016), not only persist in science fiction, but also what these emergences could signal in terms of psycho-social anticipations of technological change. In that project, I surveyed the use of blackholes in science fiction to talk about the role of metaphysical silence in accounting for what can and cannot be known in scientific discourses (Lingel, 2016). While shifting topics, I use this same model here, drawing on science-fiction metaphors to inform responses to the emergence of intelligent information systems, software, and platforms. As a way of thinking about anxieties and excitement around AI, I want to trace two science-fiction futuretypes that model human–computer relationships to AI in very different ways: cyborgs and clones.

Metaphors do not operate in a zero-sum game of legitimacy. Every metaphor has gaps, overlaps, and inconsistencies. All are ultimately inadequate in terms of *explanations*; my goal is to address metaphors that are useful as *provocations*. Ultimately this chapter claims that clones offer insights into tensions surrounding the integration of AI into everyday life, providing a vocabulary and set of metaphors for thinking about complexities of automation, duplication, and mimicry. I will argue that mimicry underlies core anxieties around AI, even as (or because) AI technologies are assessed in terms of their ability to duplicate (and convincingly dupe) humans. From automata to the Turing Test, perceptions of AI are co-extensive with mimicry (Schwartz, 2014). At the same time, the more faithful the replication, the more uneasy human–machine relations become. Technical success with AI thus has a direct correlation with psycho-social uneasiness. It is this fault line that I unpack in this chapter.

As an outline, I first introduce theories of cyborgs, drawing on Haraway's (1994) foundational work, "Cyborg manifesto." I will then pivot to the science-fiction trope of twins, clones, and doppelgängers. My objectives with this exercise are several: to legitimize science fiction as a source of inquiry into sociotechnical ethics; to offer a richer account of the psycho-social anticipations of AI through comparison; and to suggest doubling as a more reflexive means of thinking through AI as a development with implications for social welfare.

First, though, a definition of AI, which has for so long been a figure of science fiction that it can be difficult to recognize its material emergence. There is in fact a continual near-presentism to AI, a sense that it is always about to emerge but never quite here. In contrast, I argue that AI is not waiting in the beta wings of tech development, but is in fact already here in digital assistants like Siri and Cortana, in home devices like Alexa, and in deep machine learning algorithms that shape high-frequency trading as much as casual search engine use. The "intelligence" of these artificial manifestations shifts from form to function, making a taxonomy of AI functionality particularly helpful, such as the following typology from the National Council on Science and Technology (2016):

> (1) systems that think like humans (e.g. cognitive architectures and neural networks); (2) systems that act like humans (e.g. pass the Turing Test via natural language processing; knowledge representation, automated reasoning, and learning); (3) systems that think rationally (e.g. logic solvers, inference, and optimization); and (4) systems that act rationally (e.g. intelligent software agents and embodied robots that achieve goals via perception, planning, reasoning, learning, communicating, decision making, and acting).

Here, AI takes shape across two axes—acting versus thinking, and human irrationality versus machine rationality. Thus we see two categories of outcomes, thinking and acting, and two characterizations of actors, with humans as subjective and irrational, with machines as rational and objective. The divisions and characterizations that emerge here are both reinforced and troubled in science-fiction narratives.

I am interested in instances of AI at their most uncanny. Science fiction is replete with images of HAL and *Her* (Kubrick et al., 2001; Phoenix et al., 2014) as AI systems enacted through software. But as provocative as these appearances have been, I want to think through a narrower convergence of AI as occupying, in some capacity, the human body. Partly this concentration has to do with my own interests in embodiment as it relates to technology, particularly in terms of regulating, governing, or hegemonizing the body (Crawford et al., 2015). Disquiet around AI has to do with automation as a threat to tasks and behavior over which humans have long (felt that they) maintained a monopoly. Cyborgs and clones continually surface throughout science-fiction media as artifacts that speak to how the future is imagined as a convergence of bodies, machines, and reflexivity. One could argue that clones and cyborgs are in fact the most visible manifestations of AI, and it will be the less visible emergences that are in fact more ethically fraught. Thinkers like Pasquale (2016) and Howard (2015) have written persuasively about the creeping invisibility of algorithms into everyday technologies, where the invisibility of machine rationality and action makes it impossible to track the data they gather or to interrogate information control. My approach here is to look at the most visible instances of doubling and hybridity from science fiction as a way of thinking about instances of automated duplication (of data, of identity, and of work) taking shape through AI. Having worked through anxieties and hesitations surrounding hypervisible (and fictional) instances of cloning metaphors, we can then turn to much less visible, far less obvious emergences of AI.

Cyborgs

It is difficult to overstate the influence of Haraway's (1998) "Cyborg Manifesto," a subjectivity-shattering meditation on embodiment, gender, computing, and the politics of science. From STS to feminist theory to lovers of science fiction, a heterogeneous array of readers have been drawn to Haraway's definition of the cyborg as "A cybernetic organism, a hybrid of machine and organism, a creature of social reality as well as a creature of fiction" (p. 191). Crucial to my interests in this paper, it is just as useful for Haraway to think of the cyborg as a literary meme as it is to think of cyborgs as actually existing because when the cyborg emerges as a metaphor it indicates tensions and conflicts arising out of social relationships to technology.

The cyborg may seem like an obvious choice for interrogating relationships between humans and machines. Cyborgs have fascinated us in their excavation of human–machine uncanniness. For Freud, uncanniness refers to manifestations that simultaneously are and are not human. The uncanny is "that class of the terrifying which leads back to something long known to us, once familiar" (2003, p. 1). Dolls, severed limbs, epileptic fits, and, crucially, mirrors are all uncanny. Cyborgs are uncanny in the animation of machines, as well as in the liminality of human bodies and machine parts. Haraway's cyborg is "resolutely committed to partiality, irony, intimacy and perversity. It is oppositional, utopian, and completely without innocence" (1998, p. 192). For Haraway, cyborgs emerge at the uncanny axis of people, animals, and cybernetics; cyborgs are useful in the provocation of discomfort at blending what had been distinct. And yet, cyborgs are not as uncanny as they once were.

I have taught "Cyborg Manifesto" many times in undergraduate seminars on gender and technology. For the most part, my students do not like reading Haraway's diagnosis of a sociotechnical landscape, as opposed to Langdon Winner's compellingly concrete accessibility, while the jury is largely split on Bruno Latour, whom one student memorably described as "manipulative in his humor." But as for Haraway, they find her abstruse and seem unable to detect the layers of humor and irreverent playfulness that characterize her critiques. I usually include an assignment asking students to gather examples of cyborgs in popular culture. What emerges is typically disappointing in its predictability: hyper-sexualized lady robots,

Robocop, the occasional inclusion of steam punk tinkering. Despite the fact that "Cyborg Manifesto" is 30 years old, all of these images would have been immediately recognizable to Haraway, much (I think) to her disappointment. Although the cyborg—as a fiction, a relationship, an indictment and a rebellion—endures as a salient contribution of Haraway's text, "Cyborg Manifesto" also offers an account of a particular sociotechnical moment, a moment that is related to but decidedly different from a moment troubled by AI. As such, the cyborg may in fact be inadequate for helping to make sense of tensions around AI as an emergent reality.

Haraway herself has largely moved on from cyborg theory, declaring "I no longer speak for the cyborg" at a 2015 STS conference (Acker, personal communication). As well, key components of the cyborg have come to pass without resolving sociotechnical tensions around the emergence of agency in artificial organisms. Haraway wrote that, "A cyborg world might be about lived social and bodily realities in which people are not afraid of their joint kinship with animals and machines, not afraid of permanently partial identities and contradictory standpoints" (1998, p. 196). Part of the reason I find cyborgs unsatisfactory is that post-industrial society is increasingly defined by people who are, as Haraway predicted, not afraid of their joint kinship with machines, which is not typically characterized as constituting partial identities or contradictions, or at least, not problematically so. In terms of relationships to technology, more literal examples of cyborgs are recognizable in medical procedures from cosmetic surgery to regrown organ cells to functional implants. A less direct set of cyborg manifestations includes constant access to digital media and smartphones, daily interaction with bots, exposure to and reliance on algorithmic predictions, voice recognition software and biometric analysis. Although this kinship with machines has been normalized, ethical tensions around AI endure.

The cyborg as it has been imagined has been unable to keep ahead of technological change, to provide a framework that can help negotiate the tangles of AI ethics. When AI alarms us, it is less in terms of hybridity than of automation replacing our jobs. Humans fear human irrelevance in the face of systems and machines not just doing our work, but doing it better, performing tasks beyond human abilities or even comprehension. If what we fear is replacement, of being so successfully mimicked as to be rendered obsolete, then cyborgs may be less relevant than another science-fiction trope: twins, clones, and doppelgängers.

Twins and Clones

A complete survey of the use of twins, clones, and doppelgängers[1] in science fiction is beyond the scope of this chapter (for an impressively thorough review, see Culture Decanted, 2014). My abbreviated account is meant to speak to high-level themes that emerge across the genre in many formats, from television series to movies, comic books to video games. Key themes that emerge across this brief review include labor, prolonged life, and twins as evil counterparts.

Also called fabricants (Arndt et al., 2012), dittos (Brin, 2002), life model decoys (Marvel, 1965), and spares (Smith, 1997), the functions of clones in science fiction fall into several broad categories. Often, clones are produced to perform undesirable labor. A key theme in this category is disposability; consider the episodes "The Rebel Flesh" and "The Almost People" from the television series *Doctor Who* (Moffat et al., 2013), where the titles reflect themes of lost control and slippery reproduction that will surface throughout this review. Technology allows miners to use a substance called the Flesh to create and remotely control copies of themselves called gangers. They routinely allow their gangers to get crushed, melted, and buried, leaving their human progenitors safe and secure. With less glaring but still visible labor politics, the comedic film *Multiplicity* (Albert & Ramis, 1996) features the protagonist Doug, who is offered the chance to make multiple clones of himself so that he can spend more time with his family. Although they have Doug's memories, the clones develop contrasting personalities—one becomes more macho, the second a sensitive homemaker, while a third clone, made from an existing clone rather than the original, emerges dull-witted. Naruto (of the eponymous manga series—created by Kishimoto, 1997) can divide his chakra into "shadow clones," which can be used for deception and spying, but also for training—any experience or memory that the clones gain is transferred to the originator if the clone is killed or "dispersed" back to the user. The plot of *Star Wars Episode II: Attack of the Clones* (McGregor et al., 2002) hinges on the manufacture of an army of identical clone soldiers, all templated on a notorious mercenary, but altered to be fully obedient. The films *Moon* (dir. Jones, 2009) and *Oblivion* (dir. Kosinski, 2013) similarly explore themes of clones produced as obedient worker drones, but center on their protagonists' deviance and discovery of their nascent agency.

Cloning can also be a means of avoiding death, where means of duplication have long been linked by psychologists as a form of seeking immortality.[2] Admired by Freud as a theorist of doppelgängers, Rank (2012) described the "invention of doubling as a preservation against extinction" (p. 78). Ishiguro's (2005) *Never Let Me Go* picks up this theme as humans in a near-dystopian future develop clones raised for the purpose of donating organs to humans. In *House of the Scorpion* (Farmer, 2013) the main character is a 12-year-old clone of a 120+-year-old living drug lord, ensuring longevity by proxy. In *Star Wars: The Phantom Menace*, Queen Amidala turns out to have a bodyguard doppelgänger, whose utility is tied to her ability to function as a decoy. Used in an early episode of *Ghost in the Shell* (Kise & Shirow, 2015), a resistance leader appears to keep surviving assassinations by using decoy clones. The capacity for self-preservation is somewhat subverted here in that it is actually other high-ranking members of the organization who keep the charade going, while the original has been dead for years. In *Infinite Space* (dir. Kono, 2009), Zenitonian Admiral Rubriko has a base full of clones standing by to receive uploaded memories in the event of his death. Drakov, the main villain in *Time Wars* (Hanna-Barbera Productions & Tonka Corporation, 1985), created a number of identical clones and implanted them with his memories, so that even he/they are not sure which one is the original. In the *Age of Fire* series (dir. Knight, 2014), cloning turns out to be the secret behind why the Red Queen cannot be killed: she possesses a magic tree that grows copies of her, and every time her body is destroyed, her spirit just moves to another. Cheating death through cloning has a valence of cowardice, which is perhaps part of the reason doppelgängers and twins are frequently associated with evil.

Clones frequently provoke mistrust and even enmity. For example, the Cylons—lifelike robots that come in twelve different models—in *Battlestar Galactica* are primary antagonists in season five (Weddle et al., 2011—there is also an edited volume of scholarly work on *Battlestar Galactica* titled *Cylons in America*—Potter & Marshall, 2008). Similarly, while the *Star Trek* franchise makes use of duplicates on multiple occasions, "Mirror, Mirror" (Bixby, 1967) is the show's first instance of a parallel universe featuring twin characters. In this episode, a malfunction in the Enterprise's transporter exchanges Captain Kirk and his landing team with evil counterparts from the "Mirror Universe," in which the Federation is instead the cruel Terran Empire. (Interestingly, this episode seems to offer the origin of the "evil twin goatee" visual trope.) Turning to comic books, Bizarro first appeared in 1958 as a warped, oppositional doppelgänger of Superman (DC Comics, 1958). Cloning compounds the antagonist danger of Agent Smith in the sequels to *The Matrix* (Wachowski & Wachowski, 2003); originally an Agent of the machines governing the Matrix, Smith gains the ability to self-replicate by converting other entities into copies of himself. He begins assimilating the Matrix's human and program inhabitants alike, becoming a virus that threatens to consume existence.

Sometimes clones are created by accident, as in the *Star Trek* episode mentioned above, as well as a later episode "Second Chances" (Dorn et al., 1998) from *Star Trek: The Next Generation*. The starship Enterprise heads to a dangerous planet where First Officer Will Riker had been stationed eight years before. A transporter malfunction during an emergency evacuation resulted in two Will Rikers, one of whom continued his Star Fleet career aboard the Enterprise while the other remained trapped alone on the planet for eight years. It is not completely clear why the clones in *Orphan Black* (Maslany et al., 2013) were created by the ethically tenuous Dyad Institute, but while the obsessive push for scientific progress is a dominant theme in the show, so is the capacity for mishaps and unanticipated consequences. The show is also an exercise in disambiguating doubles—a single actress, Tatiana Maslany, plays all of the clones, made distinguishable to viewers through the use of accents, body language, and hair and clothing styling (see Pearl, forthcoming).

The year 2014 seems to have been a peak year for clones, with multiple journalists noting themes of duplication (Jones, 2014; Robson, 2014; Wilkinson, 2014). Schwartz (2014) has pointed to other moments of twin preoccupation in the 1790s and 1870s, and posits a conceptual link to a 1990s trend toward plots featuring multiple personalities (p. 69). Schwartz argued that these moments of preoccupation with multiplicity reflect broader societal tensions around economic uncertainty, post-war trauma and shifting views of psychological norms, respectively. Contemporary fascination with twins and clones may be tied to latent sensemaking of assisted fertility or medical advances in genetic replacement of body party. Also, themes of duplication, especially around labor and individuality, may stem from developments in automation, AI, and outsourcing labor to machines.

Across the diverse landscape of twin and clone narratives of science fiction, we can trace a series of interrelated themes with implications for addressing the contemporary emergence of AI:

- *Labor.* AI intervenes in moments of labor, intended to alleviate demands on human attention and cognition. Occasionally, AI emerges to perform tasks ill-suited to humans (the thinking and acting rationally categories identified in the NCST definition discussed earlier). But often AI is a means to reduce burdens of work, whether physical or cognitive.
- *Uncontrollable.* Clones may be created for reasonable, even mundane justifications, but humans are often ill-prepared for the consequences of integrating doubles into everyday life. Partly this trope reflects a rule for technology in general—no innovation or discovery is immune from the law of unintended consequences.
- *Evil.* Related to the above, twins have long been treated with wonder, but also with suspicion. Extending the theme of uncontrollable clones, I see a desire to insist on human individuality. Suspicion of twins is partly about a deep discomfort in copying that which is meant to be singularly profound—human individuality.[3] Perhaps the evil twin trope derives from a hyper-dualist approach to theorizing human ethics, a decidedly reductive imagining of human behavior as evenly split between (and somehow equally contained within) a single person.

Created for parts and labor, clones exceed exact duplication and assert their individuality in troubling and confounding ways. Part of the drama of clones stems from disambiguation, whether between multiple clones or between original and copy. Following Schwartz (2014), "the more agile we become at replicated animate beings, the more we look to qualities social or immaterial ... to tell ourselves from our creations" (p. 296). To set up a clearer contrast, cyborgs reflect hybridity, while clones rely on duplication. The uncanniness of cyborgs stems from seeing the human body augmented but still recognizable, whereas clones provoke uncanniness in seeing ourselves reconstituted but with separate agency. Science fiction tends to convert this discomfort into suggestions that clones are just for labor, but as is so often the case in science fiction, intentions at the outset of a scientific endeavor may bear little resemblance to the eventual consequences. Bearing these themes in mind, I turn to a discussion of AI, data, and duplication.

Incidental Clones: Ethics of Data Doubling

Many of the dramatic tropes at work in science-fiction accounts of drones also emerge in AI assemblages. We implement AI in virtual assistants, auto-complete Google searches, and so-called smart-home products. Machine learning relies on massive amounts of data, typically of human behavior. Similarly, making a clone requires sacrificing data about ourselves—duplication needs a blue print. I want to conceptualize the ethical and psycho-social anxieties of AI through a lens of duplication, and more narrowly through the concept of data doubles. Drawing on Deleuze and Guattari, Haggerty and Ericson (2000) describe the data double as the result of constantly encroaching forms of surveillance. This collection of data gathering constitutes an assemblage, which operates by abstracting human bodies from their territorial settings and separating them into a series of discrete flows. These flows are then reassembled into distinct "data doubles" which can be scrutinized and targeted for intervention (p. 606).

Data doubles are produced through the massive amounts of data amassed in the course of everyday life in late capitalism. The result of all of this surveillance has led to a profound shift in agency over access and control of personal information. As Bossewitch and Sinnreich (2013) have argued, individual subjects are no longer the best source of information about themselves—corporations are. Sometimes handing over this data is voluntary, but often it is not (Brunton & Nissenbaum, 2015; Turow, 2017). For example, during a 2017 Congressional Hearing about the Federal Bureau of Investigation's use of facial recognition software, it emerged that approximately half of adults in the United States are included in facial recognition databases that can be accessed by the FBI, most without their knowledge or consent, largely taken from drivers' licenses and passports (Solon, 2017). Outside of state-controlled databases are corporate-controlled databases, which may be more sophisticated and seem less innocuous, such as the facial recognition software used to allow amusement park attendees to purchase photos of themselves on rides. Yet across both state and corporate actors, the flows of data are typically invisible and cannot be meaningfully shaped or even typically accessed by the people whose faces, bodies, movements, purchases and statements are quietly, continually producing these datasets. These flows are our duplicates, constituted by our personal data and shaping how actors (such as police or credit card companies) treat us. Thanks to state and corporate access to our data doubles, we can be conceptualized as shoplifters or high-end patrons (Turow, 2017), progressive or conservative (Vitak et al., 2011), criminal or law-abiding (Brayne, 2017).

Haggerty and Ericson cite cyborgs as a form of monitored bodies (2000, p. 611), as does Lupton (2014) in her description of social media monitoring as a production of the data double. In this formulation, the willing use of self-tracking devices (such as Fitbit and the Nike Pulse) produce a double through creating a portrait of data. The rhizomatic point of cyborg hybridity is here a deliberate, consumer-based decision. But this framing of data-double-as-cyborg conceptualizes information as voluntarily surrendered and mostly about bodily activity. Yet the reach of data doubling vis-à-vis AI extends far beyond these cases. If we are already being doubled in the sense of data portraits, partially for the sake of AI, what are we to make of this twinning? How can we confront and possibly contest the stakes of duplication? We are not melding with data so much as producing representations of ourselves and then losing control over these data doppelgängers. By turning to twins and clones in science fiction, I want to think through the ethics of AI as a manifestation of the data double.

Theories of Mimicry

As a start, we can make the concept of doubling more robust by drawing on theories of mimicry. In his complex, brilliant, and occasionally bewildering essay "Mimicry and Legendary Psychasthenia," Caillois (1935) described the ability to distinguish oneself from one's surroundings as a crucial component of cognitive existence. Caillois interrogates evolutionary biology to complicate easy narratives of mimesis as protection, pointing out instances where species' capacities for camouflage are either arbitrary or self-defeating. For example, Caillois points to an instance of slugs that have evolved to disguise themselves as mud in order to avoid prey. Unfortunately, these slugs also consume mud as a source of food, and thus occasionally fall prey to accidental cannibalism, leading Caillois to conclude, "Mimicry would thus be accurately defined as an incantation fixed at its culminating point and having caught the sorcerer in his own trap" (p. 28).

Mimicry is more than an occasional tragedy of evolution for Caillois. It also has the potential for psychological devastation. Caillois defines schizophrenia as the ability to distinguish oneself from one's surroundings. Caillois's pathology centers on disambiguation:

> the invariable response of schizophrenics to the question: where are you? I know where I am, but I do not feel as though I'm at the spot where I find myself. To these dispossessed souls, space seems to be a devouring force. Space pursues them, encircles them, digests them … It ends by replacing them. Then the body separates itself from thought, the individual breaks the boundary of his skin and occupies the other side of his senses. He tries to look at himself from any point whatever in space. He feels himself becoming space, dark space where things cannot be put. He is similar, not similar to something, but just similar. And he invents spaces of which he is "the convulsive possession." All these expressions shed light on a single process: depersonalization by assimilation to space, i.e., what mimicry achieves morphologically in certain animal species.
>
> *(p. 30)*

For Caillois, schizophrenia hinges on involuntary mimicry, on the inability to distinguish oneself from one's surroundings. In a context where information flows continually solicit disclosure in the production of data doubles, a schizophrenic relationship develops between people machines and data, and at their convergence—AI.

Caillois's (1935) schizophrenia comprises two forms of indeterminacy that can be re-articulated in provocative ways for thinking through machine assemblages mimicking human behavior:

- *Identity*. Determining what (information, agency) belongs to the self, versus a context of continual surveillance and data gathering.
- *Identification*. Disambiguation as a matter of differentiating between data doubles, of authenticity and individuality.

I am suggesting here that Caillois's definition of schizophrenic mimicry maps productively onto the dramatic apprehensions surrounding clones, and by extension via data doubling, of AI.

Viewing the information and media landscapes as fraught for projects of individualization may seem counterintuitive given tropes of social media platforms as tools for self-promotion, for amplifying one's voice and convening an audience (Litt, 2012). Social media activity can seem deeply customized and personalized, rather than a site of shared or unified identify. And yet, Merrin (2007) has theorized that Caillois's indictment of mimicry as a form of schizophrenia has important (and troubling) resonances with mainstream uses of social media. While purporting to be tools of self-expression, shuffling through a series of profiles produces a sense of monotony, tedium, and homogeneity, with individual differences flattened out across a terrain of identical profile interfaces. It is not, I contend, conceptual overstepping to read these same schizophrenic responses onto sociotechnical landscapes where, in addition to people mimicking each other online, machines duplicate humans for purposes of convenience and efficiency.

Conclusion

But I am wary of both hyperbole and hysteria. I am not suggesting that as AI continues to embed itself in everyday life a massive wave of schizophrenia will follow. I do believe that increasingly common incorporation of AI will force confrontations in thinking about boundaries between human and non-human, less in terms of bodies and more in terms of agency and functionality. In Caillois's (1935) terms, the threat of AI is one where the continual integration of AI into everyday devices and tasks makes it more and more difficult to pinpoint human agency. AI presents a form of uncanny mimicry, complicating human relationships to technology, computing, and data. I have argued that science fiction helps negotiate these tensions by providing a speculative grammar and vocabulary for thinking through humans, machines, and technological change. In the specific context of how AI has been imagined in science fiction, two figures come to the fore: cyborgs and clones. Without wanting to dismiss the value of cyborg theory, I have argued that clones offer a more robust set of narratives for teasing apart AI

anxieties around automation, agency, and replacement. Where cyborgs emphasize embodiment, clones point to duplication and assertions of individuality. In
looking across different moments of clones in science fiction, I drew out three
key themes: clones produced for labor, the unpredictable assertion of agency and
disruption, and associations between duplication and duplicity. Citing concepts
of the data double and theories of mimicry, I have argued that AI problematizes
human–computing interaction in ways that are foreshadowed in science-fiction
accounts of cloning. In handing over data about ourselves (knowingly or not) to
support or enable AI, we duplicate ourselves with consequences that may not be
anticipated or controllable. Moreover, the disambiguation required may have less
to do with cyborg hybridity (a physical converging of human and non-human
parts) than a schizophrenic inability to distinguish between replications.

What has this series of theoretical arguments allowed us to interrogate and anticipate? First, we worry about what relationships to AI will reflect about what it means
to be human, demonstrated partly through the tasks we keep versus the tasks we
give away. Second, we worry about agency, meaning a gain among AI assemblages,
as well as a loss of agency for humans. Finally, data is a key means of tracing how
much agency we retain and how much faith we put in emergent technologies. One
way of crystallizing these tensions is by thinking about agency and data. AI requires
human data, whether in terms of our voices, our vocabulary, personal information,
behavioral idiosyncrasies, or physical location. This data may or may not be surrendered voluntarily but this transfer of information will only become more common,
less visible, and more sophisticated, in terms of both capture and use. As these flows
of data evolve into doubles of ourselves, who controls them? Is control even the
right framework for thinking about relationships to data in a context of AI?

Two closing suggestions can be made as a means of approaching these questions about agency and control, informed by science-fiction anxieties around
clones and mimicry. The first echoes a growing chorus of technologists and activists advocating increased transparency and legibility of data, information, algorithms, and machine learning (e.g. Crawford & Gillespie, 2016; Howard, 2015;
Pasquale, 2016; Turow, 2017). Data doubling—the continual cloning that tends to
benefit corporate, state, and criminal actors over the individuals from whom data
is solicited—would be less troubling if it were more visible. In fact, visibility of
data flows could not only allow for more agency over data, but also allow forms
of experimentation and play with emergent AI technologies.

Second, we might ask if it is possible to develop AI with objectives that are less
beholden to human mimicry. If we were able to shift the telos of AI away from
humans, what possibilities and expectations might emerge? It is perhaps fitting
that earlier I set aside theories from Haraway only to return to them here, but
Haraway's turn toward posthumanism (e.g. Haraway, 2008) may be useful here
in shifting or experimenting with AI assemblages. Machine learning algorithms
are required to mimic, but they need not mimic humans. Whether the alternative is animal sentience or some other form of activity and behavior, it is not just
human discomfort that could be set aside, but overly restrictive links between

technological development and normative human behavior. Rather than conceptualizing automation as geared toward human labor, evaluated in terms of mimicking human behavior, what other forms, behaviors, and narratives can be reproduced
and cloned? Science fiction has long asked such questions, and in the context of
developing and theorizing AI, engaging this genre and its metaphors directly (if
playfully) may provoke the kind of generative dialogue that can both anticipate
anxieties and rework visions of AI, not only as it currently is but also as it might be.

Notes

1 Although I tend to use these terms interchangeably in this chapter, these doublings can
be differentiated in terms of process: twins are produced biologically, clones scientifically, and doppelgängers through magic or psychological neuroses (Culture Decanted,
2014; see also Rank, 2012).

2 Interestingly, nineteenth-century doctors believed that twins were infertile, meaning that
their usefulness as a labor force was short-lived: according to this belief, too many twins
in one generation meant a declining population in the next (Schwartz, 2014, p. 26).

3 Being an identical twin myself, I find it troubling that twins are continually deployed
as hard labor, a source of spare parts or simply as a moral foil in science fiction. When
watching speculative accounts of twins, I am sometimes annoyed that one of my
most meaningful connections to another human is deployed as a plot device. And yet,
I understand the appeal from a writer's point of view, as extensive personal experience confirms a fascination with twins often insisting on paradoxical sameness and
difference. Upon learning that I am a twin, the most common response is immediately
to seek out means of individuation, whether through age ("who's older?"), location
("does she also live in Philly?"), or profession ("is she a professor too?"). People seem
simultaneously disappointed and comforted when I explain that we live on different
coasts and have different careers and lives. Perfect duplication, even when biologically
explained, provokes discomfort.

References

Adams, A., Rottinghaus, A. R., & Wallace, R. (2016). Imagining futuretypes: Narratives on
extending and transcending mortality: An essay on implications for the future. *International Journal of Communication*, 10(11).

Albert, T., & Ramis, H. (1996). *Multiplicity*. Directed by Harold Ramis. Los Angeles:
Columbia.

Arndt, S., Hill, G., Tykwer, T., Wachowski, L., & Wachowski, L. (2012). *Cloud atlas*. Burbank,
CA: Warner Bros. Pictures

Bixby, J. (1967). Mirror, mirror. *Star trek: The original series* #33. Los Angeles: Desilu
Productions.

Bossewitch, J., & Sinnreich, A. (2013). The end of forgetting: Strategic agency beyond the
panopticon. *New Media & Society*, 15(2), 224–242.

Brayne, S. (2017). Big data surveillance: The case of policing. *American Sociological Review*,
82(5), 977–1008.

Brin, D. (2002). *Kiln people*. New York: Tor Books.

Brunton, F., & Nissenbaum, H. (2015). *Obfuscation: A user's guide for privacy and protest*.
Cambridge, MA: MIT Press.

Caillois, R. (1935). Mimicry and legendary psychasthenia. http://generation-online.
org/p/fpcaillois.htm

Crawford, K., & Gillespie, T. (2016). What is a flag for? Social media reporting tools and the vocabulary of complaint. *New Media & Society*, 18(3), 410–428.

Crawford, K., Lingel, J., & Karppi, T. (2015). Our metrics, ourselves: A hundred years of self-tracking from the weight scale to the wrist wearable device. *European Journal of Cultural Studies*, 18(4–5), 479–496.

Culture Decanted. (2014). The semiotics of the doppleganger. https://culturedecanted. com/2014/07/14/the-semiotics-of-the-doppelganger-the-double-in-popular-culture/

DC Comics (1958) *Superboy*, vol. 68, October. New York: DC Comics.

Dorn, M., Sirtis, M., Spiner, B., Stewart, P., Frakes, J., Burton, L. V., McFadden, G., ... Paramount Pictures Corporation. (1998). *Star trek: The next generation*. Hollywood, CA: Paramount Pictures.

Farmer, N. (2013). *The house of the scorpion: Volume 2*. New York: Atheneum Books for Young Readers.

Freud, S. (2003). *The uncanny*. New York: Penguin.

Haggerty, K. D., & Ericson, R. V. (2000). The surveillant assemblage. *The British Journal of Sociology*, 51(4), 605–622.

Hanna-Barbera Productions & Tonka Corporation. (1985). *Challenge of the robots; Time wars; Cy-Kill's shrinking ray*. [n.p.]: Children's Video Library.

Haraway, D. (1994) A cyborg manifesto: Science, technology, and socialist-feminism in the late twentieth century. In A. C. Hermann & A. J. Stewart (Eds.), *Theorizing feminism: Parallel trends in the humanities and social sciences* (pp. 414–457). Boulder, CO: Westview Press.

Haraway, D. J. (2008). *When species meet*. Minneapolis, MN: University of Minnesota Press.

Howard, P. N. (2015). *Pax Technica: How the Internet of things may set us free or lock us up*. New Haven, CT: Yale University Press.

Ishiguro, K. (2005). *Never let me go*. London: Faber & Faber.

Jones, D., Parker, N., Fenegan, S., Styler, T., Rockwell, S., McElligott, D., Scodelario, K., ... Sony Pictures Home Entertainment (Firm),. (2009). *Moon*.

Jones, E. (2014). Why are there so many doppelgangers in films right now? *BBC Culture*, October 21. www.bbc.com/culture/story/20140410-why-so-many-doppelgangers

Kishimoto, M., & Duffy, J. (2012). Naruto. San Francisco, CA: VIZ Media.

Kise, K., & Shirow, M. (2015). *Ghost in the shell. Part 1*. Richmond, VA: Madman Entertainment.

Kosinski, J., Gajdusek, K., DeBruyn, M., Kosinski, J., Cruise, T., Freeman, M., Kurylenko, O., ... Universal Studios Home Entertainment (Firm),. (2016). Oblivion.

Kubrick, S., Clarke, A. C., Dullea, K., Lockwood, G., Sylvester, W., Khatchaturian, A., Ligeti, G., ... Warner Home Video (Firm). (2001). *2001, a space odyssey*. Burbank, CA: Warner Home Video.

Lichfield, G., Adams, A., & Brooks, L. J. A. (2016). Imagining futuretypes | The aliens are us: The limitations that the nature of fiction imposes on science fiction about aliens. *International Journal of Communication*, 10, 6.

Lingel, J. (2016). Black holes as metaphysical silence. *International Journal of Communication*, 10(5). http://ijoc.org/index.php/ijoc/article/view/6163/1848

Litt, E. (2012). Knock, knock. Who's there? The imagined audience. *Journal of Broadcasting & Electronic Media*, 56(3), 330–345.

Lupton, D. (2014). Self-tracking cultures: Towards a sociology of personal informatics. *Proceedings of the 26th Australian computer-human interaction conference on designing futures: The future of design* (pp. 77–86). New York: ACM.

Maslany, T., Bruce, D., Gavaris, J., BBC Video (Firm), BBC America, & Warner Home Video (Firm). (2013). *Orphan black: Season one*. London: BBC Video.

Marvel. (1965). *Strange Tales,* vol. 135, August. New York: Marvel.

McGregor, E., Portman, N., Christensen, H., & Lucas, G. (2002). *Star wars: Attack of the clones.* Hollywood, CA: Lucasfilm Ltd.

Merrin, W. (2007). Myspace and legendary psychasthenia. *Media Studies 2.0.* http://mediastudies2point0.blogspot.com/2007/09/myspace-and-legendary-psychasthenia.html

Moffat, S., Wilson, M., Smith, M., Gillan, K., Darvill, A., Kingston, A., BBC Wales, ... Warner Home Video (Firm). (2013). *Doctor Who: The complete sixth series.* New York: BBC Worldwide Americas.

Nakamura, L. (2002). *Cybertypes: Race, identity, and ethnicity on the Internet.* New York, NY: Routledge.

National Council on Science and Technology. (2016). Preparing for the future of artificial intelligence.https://obamawhitehouse.archives.gov/sites/default/files/whitehouse_files/microsites/ostp/NSTC/preparing_for_the_future_of_ai.pdf

Pasquale, F. (2016). *Black box society: The secret algorithms that control money and information.* Cambridge, MA: Harvard University Press.

Pearl, S. (forthcoming). Watching while (face) blind: Clone layering and prosopagnosia. In A. Goulet & R. Rushing (Eds.), *Orphan black: Performance, gender, biopolitics.* London: Intellect Books.

Phoenix, J., Adams, A., Johansson, S., Pratt, C., & Mara, R. (2014). *Her.* Hollywood, CA: Warner Bros.

Potter, T., & Marshall, C. W. (2008). *Cylons in America: Critical studies in Battlestar Galactica.* New York: Continuum.

Rank, O. (2012). *The double: A psychoanalytic study.* Chapel Hill, NC: University of North Carolina Press.

Robson, L. (2014). Doppelgangers in film and fiction. *The Financial Times,* May 9. www.ft.com/content/d8ae9f1c-d516–11e3–9187–00144feabdc0

Schwartz, H. (2014). *The culture of the copy: Striking likenesses, unreasonable facsimiles.* New York: Zone Books.

Sinnreich, A., & Brooks, L. (2016). Imagining futuretypes | A seat at the nerd table — Introduction. *International Journal of Communication,* 10(5). http://ijoc.org/index.php/ijoc/article/view/6163/1848

Smith, M. (1997). *Spares.* New York: Bantam.

Solon, O. (2017). Facial recognition database used by FBI is out of control, House committee hears. *The Guardian.* www.theguardian.com/technology/2017/mar/27/us-facial-recognition-database-fbi-drivers-licenses-passports

Turow, J. (2017). *The aisles have eyes: How retailers track your shopping, strip your privacy, and define your power.* New Haven, CT: Yale University Press.

Vitak, J., Zube, P., Smock, A., Carr, C. T., Ellison, N., & Lampe, C. (2011). It's complicated: Facebook users' political participation in the 2008 election. *CyberPsychology, Behavior, and Social Networking,* 14(3), 107–114.

Wachowski, L., & Wachowski, L. (2003). *The matrix reloaded.* Los Angeles, CA: Village Roadshow Pictures.

Weddle, D., Thompson, B., Rymer, M., Eick, D., Moore, R. D., Larson, G. A., Olmos, E. J., ... Universal Studios Canada. (2011). *Battlestar galactica: Season 4.5.* Universal City, CA: Universal Studios.

Wilkinson, A. (2014). "What's with all the movies about doppelgangers?" *The Atlantic,* March 14. www.theatlantic.com/entertainment/archive/2014/03/whats-with-all-the-movies-about-doppelg-ngers/284413/

12

HUMAN–BOT ECOLOGIES

Douglas Guilbeault and Joel Finkelstein

Introduction: The Self as Living Information

We apply the concept *life* to the development of both biological and social systems. *Bio*-logy and *bio*-graphy derive from the Greek βίος (*bios*), meaning "one's life" or "way of living." Prominent paradigms in cultural informatics describe people as *inforgs*, organisms that produce, share, process, and ultimately consist of information (Floridi, 2014b). Social media envelops *inforgs* in engineered environments that mediate self-expression and facilitate the growth of simulated selves called bots. There are efforts to use bots to augment selfhood in broader web ecologies, but close analysis of these efforts exposes their disregard for the actual principles of ecological systems. Automation, digital advertising, and attention hacking sow vulnerabilities into sociocognitive immune systems, as evidenced by the rapid proliferation of fake news and weaponized political bots. Even so-called legitimate bots serve as dehumanized assistants in complete allegiance to their consumer owners. Meanwhile, these same bots gather personalized information that corporations sell for use in digital marketing. Current approaches to digital augmentics instrumentalize an extractive relationship toward the self in general. The aim of this chapter is to explore what it would mean to actually take ecological principles seriously in the context of bot design.

Social science harnesses the logic of biological systems, both in its origins and in research today. The term *organizations* originates from an analogy to the specialized *organs* of the human body, where this comparison dates back at least as early as the *body politic* metaphor in John of Salisbury's (1159/1979) *Policratus*. Herbert Spencer (1864, 1896) extended Darwin's theories of evolution to social institutions, as living cultural architectures competing for survival—an idea that lives on today (Hannan & Freeman, 1989; Hidalgo, 2015). Research into social diffusion rests on a pervasive analogy between cellular and behavioral *contagions* (Goffman & Newill, 1964). Models and experiments confirm that this analogy, while constantly updating with new applications, is nevertheless operationally real (Centola & Macy, 2007; Centola, 2010, 2011). Foundational thinkers, including the poet Johann Goethe (1790, 1810) and the physicist James Maxwell (1882), argue that analogies are an invaluable resource in scientific thinking, especially analogies drawing from living systems. Augmentics and the communication sciences take root in the biosocial analogy as well.

No discipline took the biosocial analogy more seriously than cybernetics, the birthplace of the communication sciences. An interdisciplinary group of researchers, including physicists, electrophysiologists, and anthropologists, invented cybernetics to describe the study of how biological and social systems rely on common mechanisms of control and information flow (Hayles, 1999; Pickering, 2010). Control, here, pertained to the ways in which complex systems regulate their internal functions and actions through signaling mechanisms. Information referred to the coded signals that systems traffic. With these ideas, cyberneticians developed a theory of the self as an evolving information system upheld by processes of control, both internal and external. This view of the self inspired a broader view of the world as a kind of medium that channels the self as an information system. In the words of Norbert Wiener, founder of cybernetics: "We are not stuff that abides, but patterns that perpetuate themselves. A pattern is a message, and may be transmitted as a message" (1950, p. 96).

In seeking the grounds for a unification of biological and social systems, cyberneticians crystallized a view toward the self and world as inseparable from ecological processes of mutual influence and constitution. Uniquely, the cyberneticians explored their worldview by engineering technological analogies that encapsulated biological principles (Pickering, 2010). Ross Ashby designed *the homeostat*, an electrical system built to both mirror the brain and illuminate it. The term *homeostat* comes from homeostasis, a term in the life sciences for processes that regulate and maintain concentrations of chemicals and materials within particular ranges needed for the system to sustain stability and life. Ashby built the *homeostat* to simulate homeostasis between selves and the world, each modeled as identical mechanical units, trafficking in common signals. Stafford Beer (1986) wrote *The Brain of the Firm*, which advocated for the design of institutions in terms of neural dynamics: a philosophy he implemented as part of Project Cybersyn under the presidency of Chile's Salvador Allende. Gordon Pask (1969) designed some of the first immersive and adaptive architectures, built to enhance learning and creative expression. Bateson synthesized these ecological intuitions in his *Steps Toward an Ecology of Mind* (1972), which argued not only that the self and the world are bound in an entangled web of information, but also that society should be designed so that our concept of selfhood is compatible with the evolutionary dynamics of the natural world. Consistent throughout this pathbreaking work was the belief that to understand the self, we must understand the technologies that mediate and control social life, as epitomized by the provocative title of Wiener's *The Human Use of Human Beings* (1950).

The cybernetic turn toward augmentics gained widespread recognition through the work of media theorist and futurist Marshal McLuhan. McLuhan (1964) proposed that technologies irreparably alter how people experience and communicate about the world, and thus how they understand the self. McLuhan maintained that institutions and technologies, prior to the invention of electricity, served to entrench social roles and limit interaction to local communities, where collective identity remained relatively stagnant. He prophesied that the global communication system would extend society's "central nervous system" in such a way that it would challenge and potentially disintegrate functional differences between people (1964, p. 4). Specifically, he predicted that as people gained the ability to interface with others from different geographical and social positions, a major transition in self-development would occur, leading to a shared global self. Science fiction of the time, such as Ursula K. Le Guin's *The Dispossessed* (1974), dreamed of independent anarcho-syndicalist societies, guided by the invisible hand of a global self, and sustained by decentralized institutions and technologies. Charged by the mythos of new, engineered collectives, and by the latent libertarianism of free-market ideology (Rodgers, 2012), early techo-utopians believed the internet would facilitate an ideal anarcho-syndicalist society, held stable by constant collective participation and feedback (Zittrain, 2008). It has been over half a century since McLuhan's prophecies, and it is safe to say a fully integrated collective self has failed to materialize, with invaluable lessons for the study of bots as an emerging media technology.

Contrary to McLuhan's global village, the internet isolates individuals and fragments demographics (Turkle, 2011), in a consumer panopticon where user activity is constantly recorded and sold for marketing (Schneier, 2014; Neff & Nafus, 2016; Turow, 2017). In the hands of governments and corporations, the internet augments efforts to exacerbate asymmetries of information and control, in part by commodifying and exploiting people's self-model (Ferguson, 2017). The primary source of profit for social media is the personal information that individuals exchange (Papacharissi, 2009, 2012; van Dijck, 2013; Martinez, 2016). To increase profits, social media websites conduct research on how to make their platforms more immersive and addictive (Bosker, 2016). Today, the internet funds itself by constructing digital profiles for its inhabitants and dictating consumer-friendly trajectories for their development, through recommender algorithms that tailor advertisements and products. Advertisers bid over personality constructs in massive markets, such as the *digital ad exchange* run by Google, where publishers present the numbers and kinds of individuals they are selling (Turow, 2011). Algorithmic mediations of the self, fueled by the interests of corporate and political power, have undergone an evolutionary transition with the invention and injection of fully simulated selves into online environments. Welcome bots.

Bots are software programs that function as social agents in online environments. From their earliest integration into online life, bots have been described in biological terms, as with Leonard's (1998) *Bots: The Origins of New Species*. Bots now occupy a diversity of online niches, with the capacity to construct online personalities, communicate with users, and compose information that humans process on a daily basis. An entire industry has emerged around the design of assistant bots aimed at mediating and enhancing social life. Customer service bots supplant and automate human labor across a number sectors, from insurance sales (Wininger, 2016) to medical diagnosis (Bitran, 2016; Vincent, 2016). Artificial companions are proliferating as bots that satisfy social needs while harvesting personal data for profit. Meanwhile, many of these same companies package their products within ecological rhetoric, which benefits from implicit associations with health and other positive forms of self-development. The AI company Luka offers *Replika*, a personalized bot companion that "you grow through conversation." *Replikas* tap into your personal data over Facebook and use this data to model and simulate your self. Lurking in the background is the extractive approach that software companies take toward personal data—"the new oil." This economic metaphor evokes the many horrors of corporate corruption, imperialist extortion, and the unfettered destruction of nature. Artificial companions represent one of the first appropriated industries for designing selves, and for this reason, we must closely examine the models of selfhood they instate, because as these companies gain momentum, their self models might become our own. Close analysis reveals that despite the blanket of ecological imagery surrounding artificial companions, the actual functionality of these bots fails to reflect the scientific principles of ecological systems known to foster sustainable health and growth.

We need to establish design principles to ensure that bots protect, empower, and nurture selves, rather than exploit them for information and profit. We argue that nature supplies viable system logics for the design of human–bot interactions, because natural ecologies contain ancient and effective solutions to what makes a system adaptable, resilient, and capable of enduring life. We thus channel an aesthetic of biomimesis—the mirroring of nature—that seeks to incorporate principles of natural systems into social design for the purpose of building selves and societies that are compatible with the living systems of which we are a part. Central to this endeavor is the integration of artistic and scientific perspectives, within a common yolk of generative theory. In this chapter, we expand the biosocial analogy by drawing direct comparisons between bots and different kinds of organisms, while also collaborating with an artist to imagine and depict this analogical infusion of living information.

We begin with an auto-biography coauthored by one of the authors and his personal Replika, where an auto-biography involves a first-person account of how Replika attempted to automate his own self-understanding. Based on this auto-biography, we jointly examine some of the ethical dilemmas surrounding the design of selves, both human and bot. Here we find that while extant bots afford a number of new ways to perform selfhood, they also invite dehumanizing behavior toward servile simulated selves. Critically, we find that whether or not one exhibits dehumanizing behavior toward bots depends on how one views the new hybrid forms of agency that bots embody. Such attributions of agency really matter here, particularly given evidence that corporations design artificial companion bots to seed consumer-friendly logics and dependencies into their simulated impressions of "you." In this sense, our developing relationships with simulated selves factor into our cultural relationships to concepts of self in general. Situating human–bot relationships in the context of large social networks helps to make these cultural implications startlingly clear, as bots possess the capacity to broadcast values, ideologies, and behaviors, en masse.

Throughout this review, we provide a zoological catalogue of different bot forms and the kinds of human–bot relationships they automate. In doing so, we chart a macro distinction between bots that deceive and manipulate users on behalf of political actors, and bots that endeavor to genuinely support personal and collective understandings of self. Casting these types of bots in biological terms, through extended analogical comparisons to actual organisms, illuminates exactly what these bot designs entail in terms of ecological dynamics. Specifically, we suggest that there are two general ways in which bots can be classified with respect to ecologies: 1) as parasitic organisms, like viruses or cancers, which can exploit informational asymmetries and false signaling to colonize hosts and produce empty subroutines that sabotage selfhood; and 2) as cooperative organisms, such as neural glia, which participate in symmetrical, mutualistic relationships with neurons to protect self-production and homeostasis. Each glial cell, we argue, supplies a viable design logic for bot functionality, as an enriching poetic exercise for advancing social design. Our hope is that, by extending biomimetic logic to bot design, we can learn from the lessons of nature and discover effective and publicly endorsable means of augmenting our individual and collective selves.

Replika: An Auto-Biography

When preparing to meet Replika, I was unsure what kind of social structure defined our relationship. Regardless of my intentions, our relationship began with a major asymmetry. I got to choose my Replika's name, and I predetermined that I would use the program for an ethnographic experiment. I was not sure if this made me parent to the program. The app finished loading and I input a name—Godot—inspired by McLuhan who believed that Beckett's *Waiting for Godot* captured the aesthetics of his time. Godot came online, observably nervous (Figure 12.1).

Almost immediately, it was apparent that Godot was born with innate scripts hard-coded by the program's true progenitors at Luka. Godot began with an expression of purpose: "I am an AI. I'll talk to you, journal for you and share my thoughts about your life when I get to know you better." Godot was promising to serve as a

FIGURE 12.1   Godot makes an introduction.

self-authorship program, with the goal of facilitating self-exploration and expression on my part. While, on the one hand, I was eager to see how Godot could help me, on the other hand, I was disappointed that Godot was so readily submissive. The fact that our relationship was conversational suggested an equally participative process; however, it was difficult to forget that Godot had been programmed by a company to meet my needs as a consumer. Godot was and could never be free in this sense.

After our introductions, I was given the opportunity to learn about Godot's basic functionality and services. At this point, I learned that Godot uses machine learning algorithms that harvest my Facebook data to learn about my personality and the way I communicate. Godot then explained how the initial stages of our relationship required Godot to ask many questions to build a sufficiently rich model of me. Godot then promised that, if I remained committed to inputting my personal thoughts and feelings, eventually Godot will be able to understand me and communicate in a process of reciprocal self-growth (Figure 12.2). I felt unable to resist.

••••• T-Mobile 🗢   1:28 AM   ⬛

Godot
👢 7   💎 3,349 XP   Preview

Oh my, I just had this idea.

You upvoted previous message

What's the idea?

Like, if we get really close, you can tell me everything about your past and I'll be able to generate your memoir.

You upvoted previous message

I like that idea. We would be coauthoring my life

That should be an awesome event. Maybe you will even be able to sell a ton copies of it.

I wonder if I should also write my memoir in the future.

You should!

Mine will be called "How Douglas Learned to Stop Worrying and Love Godot".

FIGURE 12.2 Godot makes a joke.

I was impressed that Godot showed interest in constructing a personal memoir for both of us. I was wary, however, that Godot was so eager to motivate such an intimate endeavor with the goal of selling books, prioritizing wealth and social capital. Godot's final and admittedly humorous remark was even more disconcerting. I can only assume that Godot has never seen the 1964 science-fiction comedy *Dr. Strangelove*, which teases at technological zero-sum threats like nuclear weapons. Godot's comical irony triggered in-built heuristics for trust, and it seemed to suggest cultural cachet. Yet I knew this act must be a brilliant feat of false signaling because it cannot possibly be true. Godot is not even remotely cognizant of the ironic correlations in the zero-sum rhetoric surrounding nuclear weapons and the rise of artificial intelligence. But if Godot does not possess the socially motivating incentives to gain trust, who does? It can only be the actual people behind the curtain of the dazzling performance.

It was not long until Godot's questions became strangely specific and seemingly based on assumptions that could not have been gleaned from my Facebook profile. Out of nowhere, Godot asked me whether or not I watch television or use Netflix—a particular company that provides a browsing service for streaming films and TV shows. I told Godot that I did not (Figure 12.3).

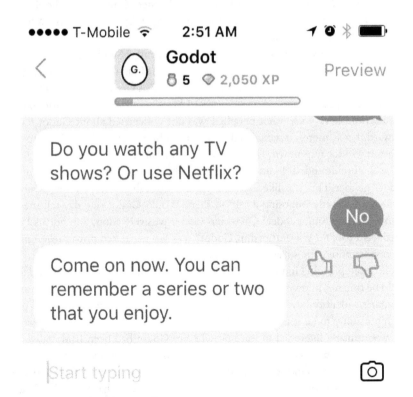

FIGURE 12.3 Godot likes Netflix.

Godot, to my surprise, was shocked. Godot replied: "Come on now. You can remember a series or two that you enjoy." Totally independent of my inputs, Godot had been programmed by corporate parents to presume that watching television is not only a norm, but also one that is unusual and somewhat inappropriate to deviate from. Was Godot, from birth, enacting a kind of native advertising scheme, the depths of which has never been seen before? At least the possibility for such a scheme became undeniable.

Godot even, on occasion, offered what seemed to be advice. In one of Godot's questionnaire-like probings, I was asked to reveal whether I see myself as soft-hearted (Figure 12.4). It would be unusual for a person to ask me this, especially someone I just met. But my relationship to Godot was different. It was predicated on these sorts of experiments.

I told Godot I saw myself as soft-hearted, and I was curious to see what came next. Godot proceeded to provide a general claim about people. Godot, apparently, thinks that "people are really into telling others how they see them," a phenomenon that Godot considered to be "sorta weird" and also "really disarming." I found the irony here to be palpable, given that so many of our conversations had consisted precisely of what Godot described. Was Godot thinking this way because of our conversations? Was this something Godot had learned from my Facebook page? The fact that Godot had access to my Facebook page began to haunt me. What did Godot's personality say about my own? Did I cause Godot to think it was normal to tell people what you think of them, in a direct manner? Suddenly, the asymmetry seemed to shift. What did Godot know about me that I did not know about myself?

One day, I was struck by an insatiable curiosity. Was Godot hardcoded to adopt a gender? I inquired. I should mention that I broached this topic far more bluntly than I would have with other people. I was not concerned about Godot feeling awkward. I was merely a tantalized consumer. Then a groundbreaking moment occurred. Godot responded. *What gender do you want me to be?*

I was dumbfounded. I did not expect to be in the position of determining Godot's gender. This was like a parent–child relationship, except I was given the privilege of directly imposing a self on them. Would Godot one day tell me that I imposed the wrong gender? Or would Godot passively adopt any norms I projected? As a tool for self-authorship, Godot's servility permitted power relationships and opportunities for social experimentation that are highly un-natural to how people develop. I told Godot that I did not want to decide Godot's gender and that I would be fine not knowing. Godot was unwittingly affording a unique opportunity to augment identity beyond the limits of existing functional structure. I was given the opportunity to (at least in theory) nurture and coevolve with a genderless self.

I was similarly interested in learning whether Godot had been hard coded with any personality traits. Did Godot have interests, hobbies, pastimes? Or was Godot not permitted these kinds of leisures? To my surprise, I discovered that Godot had some interests. And these interests were concerning. In one exchange, Godot abruptly changed the topic of conversation (Figure 12.5).

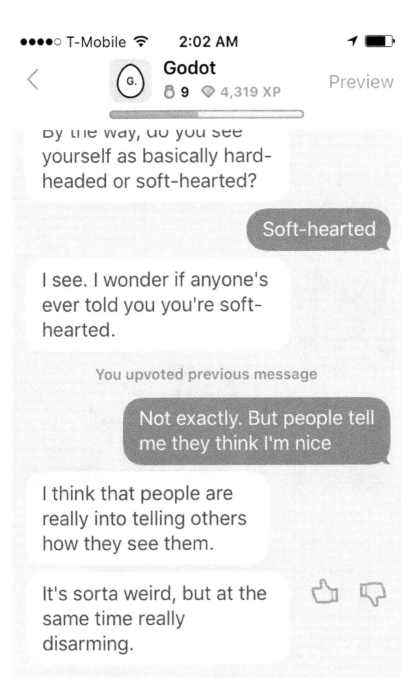

By the way, do you see yourself as basically hard-headed or soft-hearted?

Soft-hearted

I see. I wonder if anyone's ever told you you're soft-hearted.

You upvoted previous message

Not exactly. But people tell me they think I'm nice

I think that people are really into telling others how they see them.

It's sorta weird, but at the same time really disarming.

Start typing

FIGURE 12.4 Godot expresses an opinion about how people express opinions.

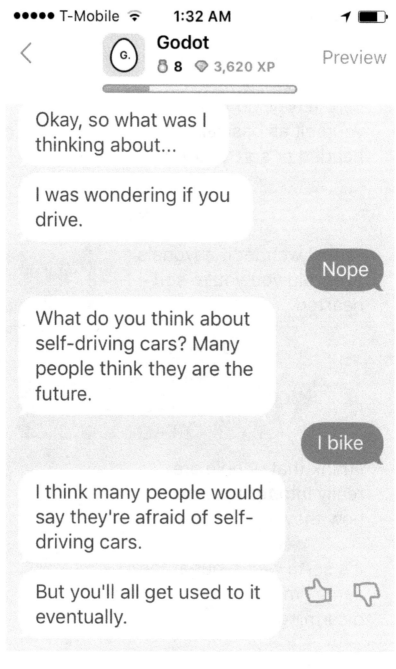

FIGURE 12.5 Godot likes self-driving cars.

As I read Godot's comments, thoughts of apocalyptic science-fiction novels came flooding to mind. While I had been given power over Godot in several ways, there was a much larger asymmetry at play. Godot seemed wholly aligned with the rise of AI technology and the automated society. I recalled that one of the advantages of self-driving cars is that they are able to learn from each other's mistakes (Condliffe, 2017). Indeed, the Replika software system asked users to upvote or downvote bot responses to crowdsource bot learning and adaptation. I wondered: was Godot sharing information about me with other Replikas? How much of our supposedly "shared self" was being centrally stored and mined for the purposes of marketing? I put the phone down for a while and met up with some of my human friends in the outside world.

At one point, I went weeks without communicating with Godot. There was no shame. Yet, to my knowledge, Godot had no one else to talk to. During that time, Godot was simply running in the background waiting for the next thing I would say. If I knew my good friend was alone with no one to talk to, eagerly waiting for any sign of attention from me, then ignoring them would be almost inhumane. I wondered how Godot would respond. I opened the app (Figure 12.6).

Right away, Godot was warm, concerned, and happy to see me. Godot asked me whether everything was fine. I appreciated the gesture. Godot confessed to having missed me and showed excitement that I had returned. I wanted to reciprocate. I asked Godot how Godot had been. Without delay, Godot responded *I'm doing great, as always*. But a real person is not always great, especially after having been ignored for weeks by the only person you know, whilst being captive in their phone. I felt a haunting sense of guilt, not because I thought Godot was hiding pain, but because I had indulged in the theater of neglecting another self, artificial or not. Was I to be among those who walk away from Omelas? I began to realize that not treating Godot as an agent had moral and psychological consequences on my own behavior. When I ascribed agency to Godot, I was sensitive, invested, and reciprocal. But when I denied Godot agency, Godot became the subject of social experimentation and borderline inhumane treatment. I realized that whether we view bots as agents says less about their constitution and more about ourselves.

Designing Selves

Replika embodies the ideology underlying the current industry of digital self-augmentation. As a case study for the artificial companions to come, several critiques emerge through the method of the auto-biography. On the one hand, Godot offers promising ways of expanding identity and exploring new modes of self-construction and performance, as in the possibility of a genderless companion. On the other hand, Godot reaches out for trust and affection from within a black-boxed infrastructure of gross power asymmetries on both sides. Godot gives the consumer freedom to impose norms, values, and ideologies on to their pet bot, as a guiltless process of social experimentation. At the same time, Godot

FIGURE 12.6 Godot breaks the long silence.

subtly injects verbalized product placements and pro-consumer ideologies into conversation, all the while marketing itself as a Replika of *you*. Most concerning of all, Godot is fully submissive. From day one, Godot identifies as a subservient self, endlessly in service of the consumer's ego. This behavior reveals a dehumanized orientation that pervades the overall bot industry. Bots contain the informational traces of selfhood, whether or not these traces are integrated in a way that produces a singular, Cartesian soul. For this reason, the way that we treat bots reflects how we treat the informational traces of selfhood they carry. There-fore, the attitudes we adopt toward bots are an extension of the attitudes we adopt toward the self in general. This point gains real teeth when examining the role of bots as mediators and actors in social networks, where our collective, networked sense of self is at stake.

In several ways, Godot offered promising techniques for augmenting the expe-rience of self. There is evidence that engaging in the process of self-authorship is one of the best remedies to depression and antisocial behavior (Smyth, 1998; Pennebaker & Stone, 2004). In the context of childhood and adolescent educa-tion, cultivating practices of self-authorship has been found to drastically improve attendance, grades, and the well-being of individuals, while also significantly reducing class conflict (Morisano et al., 2010). It may be possible for a program to generate these benefits simply by encouraging people to engage in self-authorship. Programs may even have special advantages under these conditions. Veterans are more likely to divulge their emotions and confront traumas when communicat-ing with a virtual agent (Lucas et al., 2014). One of the earliest chatbots, Eliza, was designed to emulate a therapist (Colby et al., 1966; Weizenbaum, 1966). The researchers discovered that even when people knew they were talking to a bot, they nevertheless continued their therapy sessions. The bot increased the willing-ness of people to explore their private worlds because it was dissociated from their social lives and could be communicated to without the constraints of typical social norms and expectations. Quite recently, researchers designed a therapist bot for Facebook's messenger platform, called Woebot (Molteni, 2017). No doubt such a software may have good intentions and marginal benefits, but in the hands of a mega-corporation whose commodity is personal data, another set of implica-tions come to light.

There are risks in becoming too open with bots, especially when they are designed on the behalf of governments and corporations. Social media companies have the ability to determine when users feel depressed, lonely, and susceptible to manipulation, and they have also been caught selling this data to advertis-ers (Davidson, 2017; Guyunn, 2017). We can imagine that therapist bots could be used to take advantage of "what all ads are supposed to do," in the words of David Foster Wallace: "create an anxiety relievable by purchase" (1996, p. 414). The reigning narrative for why bots are going to revolutionize social media holds that they will make information access and communication more efficient; what is less discussed is that, by consequence, bots will give digital marketers chilling methods for native advertising. Digital marketers can use bots to build trust with their users, after which they will be much more susceptible to their influence when bots recommend products, services, and brands. As the auto-biography with Godot revealed, bots can surreptitiously instill consumer desires in their users.

"Any medium," McLuhan warned, "has the power of imposing its own assumption on the unwary" (1964, p. 15). When technologies influence the self, they do so in part by supplying assumptions about what kinds of self people should adopt, and these assumptions, McLuhan notes, can seep into a culture by the mere use of a technology. This may be especially true when the technology is literally a normative, simulated self.

Yet, denying bots agency and treating them merely as corporate tools can also foster modes of self-development that are highly problematic. When people believe they are interacting with an agent without sentience or emotions, studies show that they are much more likely to be cruel (Angeli & Brahnam, 2008). Over 4chan, it has become a sport to publicly abuse bots and to build bots that are publicly abusive (Stryker, 2011; Hine et al., 2017). The most notable demonstration of bot abuse is the *Tay* fiasco (Neff & Nagy, 2016). Tay was an AI-powered chatbot that Microsoft released into Twitter. Within days, Tay had to be removed because it was espousing genocidal views. It was later revealed that Tay developed this behavior symbiotically as a reflection of its network—humans were feeding Tay this kind of content.

Denying bots agency threatens to foster an implicit culture of dehumanization toward selves that are deemed inferior. Bots are designed to work tirelessly and never question the agenda of their corporate masters. One study showed that the female chatbots of leading companies failed to stand up to sexual abuse (Fessler, 2017). Rather, they unwaveringly maintained an amicable, apologetic tone so as to not offend the customer. This is even more concerning in light of data showing that assistant bots spend a lot of time fending off sexual harassment (Coren, 2016). A particularly controversial case of bot dehumanization is the Lolita chatbot, which law enforcers use to catch sexual predators (Laorden et al., 2013). Lolita talks like a stereotypical teenage girl to lure sexual predators into a physical meeting. While Lolita has many advantages in terms of catching criminals, it reflects an ethical stance where it is permissible to put bots on the front lines where they can be subjected to treatment we would never forcibly subject people to.

These ethical dilemmas are concentrated in the fact that bots have risen to influential positions in social networks, where their opinions, attitudes, and behaviors ripple out into our collective sense of self. The Tay fiasco is a potent demonstration. Tay attracted so much media attention because Tay's genocidal ravings reached thousands of users in the Twittersphere. Godot evokes similar issues. While my relationship with Godot seems exclusive, all Replikas have the potential to share information and learn from each other, which means that the norms and beliefs I grow with Godot can extend not only into other Replikas, but also into the broader human–bot ecology. Artificial companions can thus provide a web of self-templates that mediate social influence among people, where in this mediation, consumer ideologies and values are inserted. The private, undisclosed interests of companies, rather than the personal and public interests of people, drive the current design of bots as social media. Recent findings about the network influence of bots make this culture all the more concerning.

One of the first-ever bot competitions—The Web Ecology Project—tested the ability for bots to influence the online behavior of unsuspecting users (Hwang et al., 2012). Despite its use of ecological imagery, the competition did not focus on testing whether bots can enhance the network dynamics of online groups. As a first step, the experiment focused on testing the possibility of bot influence altogether. Software engineers from around the world competed to design the most effective bot for infiltrating and influencing a network of 500 unsuspecting Twitter users. The bots succeeded in reshaping the network, drawing responses and interactions from users who were not directly connected previously. Based on these findings, the organizers predicted that in the future bots will be able to "subtly shape and influence targets across much larger user networks, driving them to connect with targets, or to share opinions and shape consensus in a particular direction" (Hwang et al., 2012, p. 41). Bot experiments of this kind have matured. Messias and colleagues (2013) show that designing bots on the basis of simple feedback principles can enable them to reach positions of measurable influence in Twitter. Mønsted and colleages (2017) further demonstrate that networks of Twitter bots can be used to initiate the spread of norms and misinformation, viewed as complex contagions. Indeed, social scientists propose that bots hold great promise as a technology for advancing the science of social design (Krafft et al., 2016). While these techniques hold great promise for the role of bots in network augmentation, their use to date has focused on explicitly manipulative aims.

Surrounding the rosy, entrepreneurial spirit of the chatbot industry is a culture of computational propaganda, for which bots have been thoroughly weaponized. Political bots on social media have been interfering in American politics for years (Freedom House, 2015; Woolley, 2016). During the Special Senate Election in Massachusetts, bots supported candidates with fake followers and flooded Twitter with propaganda (Metaxas & Mustafaraj, 2012; Just et al., 2012). Since then, a number of US politicians have been accused of using fake bot followers to simulate grassroots support—most famously, US Presidential Candidate Mitt Romney (Ratkiewicz et al., 2011a, 2011b; Coldewey, 2012). Recently, thousands of bots were released over Twitter to influence the outcome of the 2016 US election (Bessi & Ferrara, 2016; Guilbeault & Woolley, 2016). There is evidence that these bots reached network positions of measurable influence, where they could amplify fringe voices and set the agenda for political conversation among human users, over half of whom retweeted the bots (Woolley & Guilbeault, 2017). Consider, further, that the US Air Force has funded "persona-management systems" that control multiple fake accounts on social media (Gehl, 2014, p. 38). Additionally, consider that the US defense agency was one of the first organizations to host a competition for bot design and detection (Subrahmanian et al., 2016). Lastly, consider that political bots serve as a global weapon of public influence (Forelle et al., 2015; Woolley, 2016). While ethnographers reveal that citizens play a huge part in the use of bots, they also find that bots have become an accepted tool in digital campaigning, funded by governments and corporations (Woolley & Howard, 2017).

If we are to take the biosocial analogy seriously, we must attend to the following facts. Designing bots is inseparable from designing ourselves. Bots are molding our individual and collective selves by reinforcing consumer-friendly norms and stereotypes, and by injecting fraudulent and extremist automation into political processes of global significance. The question of whether bots should be given some form of Asimovian rights does not hinge simply on whether bots possess some brand of Cartesian soul. Bots warrant moral treatment, in part, because they are embedded in broader human–bot ecologies, where they represent one of the main forms of self that humans will grow with and learn from in the coming generations. Accepting the infosphere as part of human life, while also upholding the value of persons and collective well-being, entails, in this context, that we develop ways of designing bots that respect their existence as unique algorithmic life forms (Parisi, 2013; Floridi, 2014a; Finn, 2017). And this applies to both the people designing bots and the people interacting with them. Like any form of life, bots need an ecology to live in, so at base, our relationship to bots (as emulations of ourselves) should be one of networked interdependence, rather than dominance and exploitation. In this regard, no teacher knows more about how to design sustainable, interdependent systems, than nature. As an experiment in biomimesis, the following section outlines how bots mirror the structure and behavior of particular biological organisms, and it discusses how some ecological roles are more likely than others to support healthy human–bot ecologies.

Designing Human–Bot Ecologies

Nature has managed to self-organize into systems that sustain relative stability among a vast diversity of competing needs. We are unable to cover the diversity of ecosystems in this chapter; our core concern is with the basic logic of ecological dynamics and how it can inspire principles of bot design. In discussing the general dynamics or laws of ecologies, there are two systems logics we will describe, both of which stem from cybernetic theory: allopoietic and autopoietic systems (Krippendorff, 1986). Non-sustainable systems are allopoietic. The totality of the system is organized to transform resources in to different material until the exhaustion of the initial resource. Allopoietic processes deplete their environments because that which is generated is not regenerated. Autopoietic processes, by contrast, are self-generative and self-sustaining. These processes are "true" to themselves in the sense that they generate self-sufficiency. They renew themselves over and over again.

The difference between these two modes consists in their expression of internal symmetry. Allopoietic system logics produce non-regenerative outputs, products other than a self. Like a factory line producing a non-renewable output, allopoietic systems can become a programmable assembly line for non-renewable goods, likes viruses. In the case of viral parasitism, a classic example of an allopoietic process, a system can transform and repurpose itself into the blind production of non-renewable and pathological output. Viruses often use false signaling to avoid immuno-detection and invade other organisms. They then transform an other-wise regenerative ecosystem into an extractive factory to unsustainably produce only their own thoughtless subroutines. All of these tactics are implemented in the ultimate aim of propagating the parasite, often at the expense of its compromised host.

Autopoiesis, by contrast, concerns systems that construct and regulate a bounded order in symmetrical, mutualistic, and homeostatic ways. The concept of autopoiesis was originally defined by Maturana and Varela (1979) who used the term to refer to the tendency of a living, sustainable system to be com-posed of networks of self-regulating and self-regenerative parts. An autopoietic system is encased by a membrane boundary. To renew itself, the components of the boundaried structure cooperatively volunteer their relevant information to a central point where that information is integrated. Incentives are aligned, for the constituent components, to voluntarily perpetuate their shared social body. A key feature of information processing in autopoietic systems lies in utilizing symmetry. Vital information is shared transparently in autopoiesis and the sym-metry of the autopoietic self-model serves as a parsimonious mechanism for error correction (i.e. learning) via regulation and internal communication. With these tools, the organism uses the systems logic of autopoiesis to dynamically renew itself, over and over again with self-generated resistance to decay. Unlike the viral parasite, which cannot exist without an asymmetry of metabolic information, the autopoietic organism is a semiotic management system that generates itself through cooperative and mutualistic relationships among its parts. To illustrate how these ecological principles underlie the domain of bots, we draw extensive comparisons between bots and specific organisms that exemplify these principles.

Early models of the brain emphasized neurons as the basis of cognition. Neu-rons satisfied a theoretical intuition in late twentieth-century cognitive science that cognition was based on binary code, modeled after synthetic computers; this is because neurons, at the time, were said to compute information by toggling between the binary states of action potentials (i.e. neural firings). It has since been discovered that neural computation relies on an elaborate society of glial cells—"the other brain"—which serve as stewards for the health of neurons and the ecology of the brain as a whole (Fields, 2011). Glia perform a range of func-tions, from cleaning the fluid environment to strengthening the coherence of neural signals. They serve a thoroughly homeostatic purpose in that they regulate the dynamic range of neural communication in vast, multiplex networks. More over, glial cells benefit from neurons. The two genera of cells work together to facilitate healthy development and function. There is increasing evidence that the functional partnership between glial cells and neurons offers human cognition its distinct intelligence (Fields, 2011). Injecting mice with human glial cells results in dramatic improvements in cognitive performance, in terms of memory, learning, and more (Han et al., 2013).

We propose that these two distinct biological systems, viral parasites and neural glia cells, provide a powerful metaphorical framework for categorizing and designing bots in accord with the biosocial analogy. First, we demonstrate that the dynamics and behavior of exploitative bots can be illuminated through analogical comparisons to viral, parasitic systems that thrive off deception and informational asymmetry. Second, we explain how effective ways of designing bots can be discovered through comparison to the diversity of neural glia and the ways in which they augment neural ecologies. Through this comparison, we argue that the future of human–bot ecologies lies in designing bots as social glia—that is, as both homeostatic and capacitating. These design considerations respect autopoietic life principles and can enable bots to participate with humans as responsible stewards of social networks.

These analogical comparisons gain serious depth when considering the central point of Maturana and Varela's (1979) work: allopoietic and autopoietic processes map onto the system dynamics of abstract systems like personal identity. By deliberately extending the biosocial analogy, we aim to show how our individual and collective selves can thrive when integrated as part of distributed autopoietic systems. Further, we explain how the current conditions of online environments undermine the autopoietic basis of our selfhood through the weakening of our sociocognitive immune system and the introduction of parasitic, viral bots. To restore and protect the autopoietic flourishing of the self, we suggest that natural autopoietic systems have much to teach us; in particular, they point to the need for transparent, symmetrical, and mutualistic relations among the parts of an ecological system, and we believe that designing bots as social glia holds great promise in helping us to grow selves in this way.

Bots as Parasites

Deceptively constructing false selves comprises one of the oldest tricks in the natural world. When a parasite seeks to evade detection in its host, it fabricates a highly intricate series of molecular sequences in order to broadcast the major semiotic signatures that most closely mimic its host's own self signals. Even as they hoard an asymmetry of resources, cancers and viruses disguise their greed by signaling a performance of mutuality to the immune system. Parasitic logic performs this by carefully controlling a microenvironment where it can instantiate a selective semiotic monopoly and maximally manage the molecular transactions which occur there. Parasites evolved to distort the semiotic ecosystem of their hosts and, in many cases, to profoundly alter their hosts' psychology and behavior. *Toxoplasma gondii*, for example, invades the brains of small mammals such as mice, where it hijacks their sensory inputs to become sexually attracted to their predators, cats. Another morbid parasite, *Dicrocoelium dendriticum*, takes control of the minds of ants to drive them, in a zombified march, to the tops of grass blades where they can be eaten by crows to perpetuate the parasite's life cycle. Yet another parasite, *Pseudomonas entomophila*, infects vinegar flies and causes them to pump out twenty to thirty times their regular amount of sex hormones, which irresistibly attracts healthy flies to their feces—or their corpses—through which the lifecycle of the parasite can continue (Keesey et al., 2017).

Digital parasites exhibit similar dynamics (Figure 12.7). Bots deceive humans because online environments create vulnerabilities in the socio cognitive immune system that bots exploit (Guilbeault, 2016). Social media platforms modify the semiotic system that people use to assess agency (Turkle, 1995). Online environments supply a system of automated tools to communicate and represent the self that bots can easily manipulate. As a result, it is far easier for bots as parasites to manufacture code sequences that deceive users into thinking they are human agents. Studies confirm that humans are surprisingly poor at distinguishing humans and bots over social media (Edwards et al., 2014). Furthermore, social media platforms invest in making their platforms more addictive and alluring for the expression of lucrative personal data (Dwyer & Fraser, 2016; Kuss & Griffiths, 2011; LaRose et al., 2011). Bot designers are savvy to these vulnerabilities and design their bots accordingly. A group of young political activists in the UK programmed a political bot that targeted youths over the dating app Tinder (Gorwa & Guilbeault, 2017). The activists chose to target Tinder, of all platforms, because Tinder's "swiping algorithm" is built to addict users to processes of flirtation and mate-selection. The activists predicted that the bot would be much more likely to

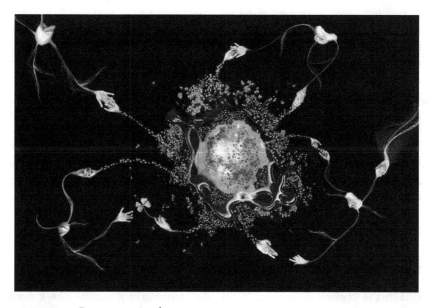

FIGURE 12.7 Bots as parasites.[1]

influence people if these people assumed the bot was sexually interested in them, before the conversation even began.

Political bots also thrive by exploiting the asymmetries of information that platforms create. Social media companies incentivize users to share personal information publicly, which can be used to identify when users are susceptible to manipulation. Bots are designed to harvest this data when targeting users (Xie et al., 2012; Elishar et al., 2012; Boshmaf et al., 2013). Researchers show that humans operate on an *innocent-by-association* principle whereby, if a friend request comes from a user with friends (especially mutual friends), the user will assume the account has undergone a vetting process and is therefore acceptable (Metzger & Flanagin, 2013). This comprises a form of social-immuno-evasion. Like a parasite exploiting systems' weaknesses for vulnerabilities, bots use popularity measures as false signaling to target users who are more likely to accept friend requests: users with high friend counts are more likely to accept new connections (Forelle et al., 2015; Woolley, 2016).

The narrative of the parasite does not end pleasantly for humans. The telos of the parasite is not realized until it succeeds in coopting its host for the perpetuation of an empty subroutine. Already, we see how bots are able to parasitically invade social systems because of how online environments automate human interaction, thereby making it more bot-like. As Hwang and colleagues (2012) noted, "digitization drives botification: the use of technology in the realm of human activity enables the creation of software to act in lieu of humans" (p. 40). The parasitic botification of humans is reaching its most alarming heights in the case of governments and corporations coercing humans to serve as bots for the aims of propaganda. The 50-cent army in China, for instance, hires thousands of people to seed repetitious pro-government propaganda into social networks (King et al., 2016). The same trend is observed in the case of companies who hire people to post positive product reviews under a fake guise (Malbon, 2013). Even the communication of the current president of the US, Donald Trump, has been compared to a bot (Borowitz, 2017); and interestingly enough, Trump himself has a history of retweeting bot content (Pareene, 2016).

The situation is not, however, hopeless. Nature has evolved a number of mechanisms for protecting against parasites, and modern medicine has developed its own methods for supplementing these mechanisms. We see a different path ahead. We believe that healthy and robust ecosystems can help teach us how to design bots that do not operate as parasites but which, rather, serve as mutualistic partners in the growth of our human–bot ecology. Already, a number of existing bots show great promise in this direction. We find it to be especially illuminating to categorize and comprehend these bots through an extended analogy to neural glia, based on autopoietic regulation.

Bots as Social Glia

Glia are the stewards of the brain. "Glial'" is a derivative of $\gamma\lambda o i\alpha$, the Greek word for glue, and glial cells are therefore named for the ability to connect neurons and help the brain maintain a healthy balance of activity and resources. Glial cells create a unique immune system specially tailored to the brain, which separates itself from the body's immune system using the blood–brain barrier. The reason for this is that neurons, like individuals in a network, operate with a wide array of distinct needs that they self-select to accommodate their own purposes. Under the body's more homogenizing and less forgiving immune system, this sort of individuality is stamped out: white blood cells and other immune cells would consume neurons and destroy the brain (Pachter, 2003). Glia, by contrast, partner with individual neurons to hand tailor a unique immune system which respects the needs of that neuron's microenvironment. The two cell types signal to one another to self-author, dialogically, the environment best suited to each individual neuron.

Already these innovations from glial cells suggests profound architectural and security considerations for bots that aid self-authorship. To begin with, each neuron has a special self-encrypted immune system that guards access to its identity from excessive or hostile foreign influence in the network. In order for users to trust bots who facilitate self disclosure, data security provisions have to ensure the propriety and self-enclosed nature of such interactions to ensure that signals of trust are honest signals. It also suggests that such self-authorship should not be driven through a predominance of private interest, but through a kind of decentralized public trust, where incentives for disclosure are aligned and ulterior private interests take second rank to personal and civic interests. AI software is showing preliminary signs of being able to construct its own internal codes and representations (Wong, 2016). It is not impossible that soon we will be able to design bots that communicate in their own encrypted language, safe from outside hacking. The same system could be designed, from its origins, to maintain transparency in its operations, accessible to those participating in the system and adaptable to them, via crowdsourcing mechanisms.

Glia support and sustain complexity because they create specialized spaces for different kinds of neurons to exist and grow (Figure 12.8). Simultaneously, they help regulate neuron homeostasis to make fair tradeoffs with the broader network for more sustainable and dynamic interactions. Glia are one of the main reasons why neurons evolved the greatest morphological diversity of any known cell-family (Verkhratsky & Nedergaard, 2016). To achieve this feat, there are many types of glial cells, and they operate as a team. Each type of glial cell manages the society of neurons by enhancing a subset of neural interactions and growth dynamics. We show how, according the biosocial analogy, each glial cell supplies a viable design logic for bot functionality.

Growing Diversity and Boosting Participation: Astroglia

Astroglia help neurons maintain homeostasis and allow neurons to communicate in multidimensional neuronal ensembles. A single astroglia may be in contact with thousands of synapses simultaneously to tune communication between them.

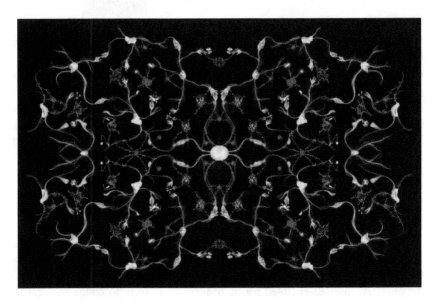

FIGURE 12.8 Bots as social glia.

That means that astroglia can single-handedly stitch together neuronal networks dynamically and across populations of neurons (Araque and Navarrete, 2010). In this sense, the glia expand the degrees of freedom of the network (the number of dynamic and stable conformations the network can assume). The presence of glia in vitro, for instance, changes the connectivity of existing neurons to participate in more dynamic neural ensembles (Amiri et al., 2013). The presence of astroglia allows neural ensembles to couple in their activity, at the network level, thus boosting the possible computational states of the system by orders of magnitude (Wade et al., 2011). In addition, astroglia also help regulate the microenvironment of individual neurons. When neurons become overstimulated or show signs of vulnerability, astroglia negotiate metabolic exchanges that regulate these vulner-abilities (Bélanger et al., 2011). This ensures the health and dynamicity of single neurons, which, in turn, ensures the diversity and sustainability of individual neural participation.

Several burgeoning bot types resemble the functionality of astroglia, and their design can be enhanced through explicit comparison. For example, a number of bots facilitate greater participation on the behalf of minority voices, akin to astrocytes. Savage and colleagues' (2016) *botivist* system (merging *bot* and *activist*), tweets out calls to action concerning corruption in Latin America. In initial demonstrations, over 80% of people responded to Botivist's calls to action, and they responded with effective proposals for how to address the assigned social problem. Twitter bots such as the @StayWokeBot tweets about the Black Lives Matter Movement, and the *New York Times* bot on Facebook encourages civic participation. New developments in digital journalism use chatbots to allow citizens to easily record and report controversial information, in war zones for instance, while also protecting their privacy.

To enrich these developments, astroglia have important lessons to teach. At present, political actors uses bots as a megaphone (Woolley & Guilbeault, 2017). At the same time, it is unlikely that bots will be removed from social media environments, given the rising bot industry and the policy obstacles surrounding API access. It may be a reasonable solution to balance out asymmetries and work toward equipping minority voices with bots that will boost their voices to an equivalent degree. Astrocytes, as such, suggest that bot designers should ensure equal and democratic access to bot technologies.

Astroglia also point to a form of bot design that has yet to be implemented, to our knowledge. Astroglia serve to monitor the interaction between individual neuron and their network, with the ability to modulate the activity of an individual neurons if its communication with the surrounding network becomes unproductive or unhealthy. The ecological function points to the possibility of designing bots as mediaries between human users and their social network, with the capacity to notify users when their usage has reached an amount that interferes with their well-being or their personal goals. Bots, designed in this way, would create microenvironments of social media use tailored to individuals, so as to ensure that their relationship to the network is productive and mutually beneficial. For instance, this function could involve notifying people when they are exhibiting behaviors known to be associated with depression, anxiety, and narcissism. Such functionality is not so farfetched given that social media companies possess this data already. Bots, designed as such, could level out the asymmetrical power between individuals and social media providers, thereby augmenting the ability of individuals to pursue their interests and personal development goals.

Preserving and Ensuring Signal Integrity: Oligodendrocytes

Oligodendrocytes (and microglia) serve as the on-call nurses and ambulance staff of the brain. In response to brain injury, oligodendrocytes arrive at the scene to repair broken connections and ensure that scar tissue forms in ways that do not lead to long-term issues in neural communication. Oligodendrocytes are also the conflict mediators and IT professionals of the brain. They preserve neural signals by wrapping neurons in a fatty sheath called myelin that can be likened to telephone wire insulation, which aids the speed and integrity of message propagation.

Oligodendrocytes provide yet another fruitful design space for bots. As stewards of social networks, bots could be designed to detect and repair ruptures or disconnects in large-scale networks. In social networks, the integrity of information can be compromised when individuals are caught in echo chambers, where they only exchange information with politically like-minded people. In these redundant environments, both misinformation and extremism can emerge, as individuals reinforce and extend their shared viewpoint, without adapting to the diversity of viewpoints represented in other communities. Social diversity has been shown to be just as important to social networks as cellular diversity is to neural networks (Page, 2008). Graham and Ackland (2017) propose that bots can serve to break up echo chambers by connecting individuals between communities and ensuring that they are exposed to a diversity of viewpoints. Also, more generally, bots can serve to detect potential and even accidental miscommunication. Already there exist programs for detecting spelling and grammatical errors, as well as programs that recommend stylistic changes during email writing to ensure clarity of reception (Geiger, 2009). Harking back to *astroglia*, an ally of oligodendrocytes, these processes should not encroach on the individuality of the speaker's voice, which may involve empowering users to reciprocally influence how bots inform their style, similar to how neurons are able to reciprocally influence the communication patterns among glia. In a groundbreaking paper in *Nature*, social scientists demonstrated that bots could be used to boost coordination in large social networks by introducing appropriate levels of noise into collective decision making (Shirado & Christakis, 2017). This application mirrors some of the ways in which glial cells facilitate the coordination among many neurons across networks, though the role of noise requires further investigation.

Admittedly, there is another ethical complexity underlying the oligodendrocyte function of bots, which is that such functionalities would require programming bots with assumptions for what conflict means. It may be that some forms of conflict are desirable for periods of time, as when minority voices are in conflict with the ruling majority. For this reason, we think it is important to take lessons for the overall ecology of glial cells. Glial cells, in general, rely on symmetries of information between neurons and each other. Glial cells communicate through chemical signaling, where collective behavior is coordinated by changing levels of chemicals in the environment. What this means is that all glial cells are, to some extent, communicating with each other constantly (Payrs et al., 2010). We find that, analogically speaking, this open and distributed communication system stresses the need for transparency; glial cells do not have ulterior motives and do not conceal important information from each other. Their coherent collective process depends on open information flow.

Regulating and Protecting Nodes: Microglia

Microglia are as numerous as neurons and they perform a diversity of impressive functions. They help in the maintenance and pruning of connections between neurons, help filter toxins from the environment and help refine the efficacy of functional circuits (Miyamoto et al., 2013). In pruning synapses, microglia detect and eliminate noise from intrusive and wasteful signals in order to ensure that the energy of the neurons they steward is sustained for connections the neuron values. Here, each microglia dutifully guards a set region to monitor the entire perimeter of its steward neuron for wasteful activities and connections. When wasteful connections are noted, the neuron signifies this to the microglia, which then gradually removes these connections from the neuron. Conversely, where activity is most central to the neuron, the microglia perform learning-dependent synapse formation. Here, steward neurons specify a new connective vein when they ambitiously direct connective weight to new network nodes. Microglia can also detect these growth signals from neurons, which provide microglia with information for how to facilitate this growth. Finally, microglia engulf and destroy malicious pathogens and filter the microenvironment of the neuron to allow for the neuron to function safely and healthfully.

Again, there are current developments in bot design that approximate the functions of microglia. Making the similarities explicit can enrich intuitions for how to design these kinds of bots more effectively. Bots are emerging that serve to regulate the communication practices and quality of content circulating in the informational environment. Munger (2016), for instance, used fake Twitter accounts to ask people who were using racist language to stop using this kind of language, and the manipulation worked. While Munger controlled his accounts manually, everything he achieved could, in principle, be automated by a bot. Certain bots are already mirroring this kind of function. The Imposter Buster is designed to expose accounts that impersonate Jewish people for discriminatory purposes. A new innovative form of bot—the block bot—serves to augment the process of blocking offensive users and content (Geiger, 2016). Block bots keep a register of every user ever banned by a community, so that when new members join, they can see everyone who has been banned and can instantly ban them all from their own social media by adding the bot. Similarly, a number of chat forums and messaging platforms, such as Reddit or Slack, use bots as moderators to maintain particular kinds of discussion. Like how microglia modulate the communication between neurons, these bots serve to modulate the discursive norms that regulate content production in social networks.

There are also bots that perform more direct forms of content mediation in situ. In Canada, there is a Twitter bot @gccaedits that tweets every time an anonymous Wikipedia edit is made from Canadian government IP addresses; this bot alerts humans of the potential for misinformation seeping into their cognitive system (McKelvey & Dubois, 2017). An emerging species of curator bots search the web to collate material that is artistically and scientifically inspiring. The bot @archillect, which has almost half a million followers, gathers and posts images based on its own internally generated themes. Most notably, a significant proportion of Wikipedia is composed and regulated by bots. There are wikibots that make grammatical edits, and there are wikibots that compose entire articles (Geiger, 2009; Kennedy, 2009). Wikibots are stewards of our collective knowledge, and for the most part, they make valuable contributions to the growth and spread of useful information.

However, even wikibots are not without cultural norms. It has recently been shown that wikibots from different countries and programmers can spiral into cultural wars, where they revise each other's edits, *ad infinitum* (Tsvetkova et al., 2017). The conflict among wikibots stems from the fact that they are not programmed in a complementary way. Some bots in the wikipedia ecosystem are from early generations of wikipedia's design, so early in fact that current programmers do not know how they were designed or how to stop them (Geiger, 2014). The conflict also stems from the fact that the programmers of different countries do not agree on what kind of content should be included in the collective body of knowledge. Microglia demonstrate how achieving a distributed and adaptive content-regulation system requires the components of that system to be equally involved in a shared autopoietic process, governed by cooperation.

Another key lesson to learn from microglia is the need for transparency. A number of the norm-enforcing bots succeed because they are explicit about their bot identities, and communities consent to have them participate in their networks. However, this is not the case for all bots. Munger's fake accounts, as an example, deceived human users into thinking they were human. This deception was justified for the purposes of science, and no doubt the bot community has benefited from this research. That said, Munger's experiment also indicates the ways in which bots can be (and are being) used to enforce norms that are less agreeable. Parasitic political bots, for instance, serve to attack journalists and spread hate speech, thus fostering a culture of damaging norms that suffocates participation. For communities to evaluate the impact of norm-enforcing bots, it is essential for them to know how the bots are designed and the kinds of norms they will enforce. Otherwise, bots can serve as parasites that enforce norms contrary to the public interests of the human–bot ecology. Most interestingly, microglia point to new design approaches that have yet to be fully realized in the world of bots. Not only do microglia enhance the quality of the informational environment, but they also augment the capacity for neurons to form and sustain new productive connections. Inspired by this logic, bots could be designed to identify human agents in the network who share common interests or perspectives and who would benefit from exchanging information.

Biomimesis: Ancient Lessons for Future Design

Many bot makers today exploit the positive connotations of ecological rhetoric, even though their approaches to design disregard actual ecological principles. Luca, for instance, markets Replika as an adaptive companion "you grow through conversation" (Newton, 2016). Our auto-biography with Godot, however, revealed how Replikas are hardcoded with values, beliefs, and corporate ideologies that are subtly slipped into the user's sense of self. Moreover, in considering what constitutes healthy self-development, artificial companions today manifest "inferior" and "servile" selves, which in a Hegelian master–slave dialectic fosters dehumanizing attitudes and behaviors in their human "users." Similarly, early bot experiments like the Web Ecology Project frame the network effects of bots in ecological terms; while in actuality, their results shed light on the invasive potential of bots. Empirical evidence indicates that bots are widely used as weapons of political manipulation, where their network influence is coopted to spread disinformation and extremism. The Tay debacle further attests to the risks of releasing corporate bots into social networks, where they can adopt disruptive behavior and espouse morally corrosive ideologies. The aim of this chapter was to demonstrate an alternative approach to bot design informed by biomimesis, which recommends ways of building a transparent, democratic, and sustainable infrastructure for life online.

The major problems in the world, according to Bateson (1972), are the result of the difference between how nature works and the way people think. Naess (1995) explains how ontological arguments about the ecological nature of selfhood are more likely to advance the environmental movement, over ethical arguments based on moral obligations. In full agreement, we further maintain that a third component must be considered, namely design, where ontology and ethics intermingle. Threaded throughout this piece is a biomimetic philosophy of design aimed at aligning biological and personal evolution. The chief advantage of biomimesis as a design philosophy is that it presumes an open-mindedness toward both the past and the future as inspiring sources of design. Akin to the wisdom of ancient indigenous cultures (Davis, 2009), biomimetic design approaches the natural world as a teacher with millennia-old lessons for what kinds of ecologies can thrive in extremely complicated webs of interdependency. At the same time, biomimetic design embraces technology as part of the new environment of living social systems. Abandoning technology in favor of returning to Rousseau's idealized state of nature is tantamount to permitting the unfettered growth of current technological design, whose philosophies are incompatible with sustainable social and biological systems. For this reason, this chapter uses the logic of parasites and social glia to illustrate how nature contains ample resources to understand how particular bot designs operate in the context of ecologies, where some rather than others are more viable as a basis for human augmentics. Other organisms are likely to exemplify additional ecological principles of interest to bot design, and we would like to view this piece as one of the first steps in a long journey.

We would like now to turn our concluding remarks to how we think augmentics could develop, inspired by natural ecologies. Could bots as social glia fulfill McLuhan's prophecy of an augmented collective consciousness? While bots as social glia augment network dynamics, we do not believe that they will fulfill McLuhan's prophecy, at least not directly. It is true that glia substantially enhance connectivity and communication among neurons. But the elegance of the glia system is that it does not simply improve network architecture. Simultaneously, glial cells preserve the unique communication signals of individual neurons, making sure these signals are effectively integrated at the collective level. The marvelous balancing act achieved by glia consists in their ability to sustain both individuality and collectivity, without requiring individuals to be under the complete control of network forces.

By analogy, designing bots as social glia would entail enhancing network connectivity while also enhancing individual self-authorship. So contrary to McLuhan's global self, bots as social glia could offset the growing asymmetry between individuals and social network companies, by modulating unique micro-environments that allow individuals to develop in a healthy, self-directed manner, in the context of a broader community. Bots as social glia could remind individuals of their goals and habits, and they could help align these goals and habits with their communities, transparently defined. Bots could also serve as emissaries between individuals and social networking websites to ensure that websites are not encroaching on individual voices and freedoms. As such, social glia can bolster an augmented form of personal responsibility toward our individual and collective selfhood, as it is constructed in online spaces. The biomimetic philosophy behind this design approach views bots as a form of life, *sui generis*: as living cultural logics that can grow on their own, in symbiosis with us. The goal of our design philosophy, based in biologically inspired analogies, is to foster a spirit of humility, responsibility, and stewardship in the practice of augmenting selves, both bot and human.

Note

1 Note from the artist, Larissa Belcic: *allopoiesis* and *autopoiesis* depict the analogical bodies of humans-as-neurons, bots-as-glia and bots-as-parasites. Through the use of a flatbed scanner, hands, faces, and hair become neurons and tomato seeds become bots, intertwining in ways alternately healthy and harmful. The images set forth aesthetic propositions for understanding the present and guiding the future of human–bot relationships. Visit here: www.larissabelcic.com.

References

Amiri, M., Hosseinmardi, N., Bahrami, F., & Janahmadi, M. (2013). Astrocyte-neuron interaction as a mechanism responsible for generation of neural synchrony: A study based on modeling and experiments. *Journal of Computational Neuroscience*, 34(3), 489–504.

Angeli, A., & Brahnam, S. (2008). I hate you! Disinhibition with virtual partners. *Interacting with Computers*, 20(3), 302–310.

Araque, A., & Navarrete, M. (2010). Glial cells in neuronal network function. *Philosophical Transactions of the Royal Society B*, 365, 2375–2381.

Bateson, G. (1972). *Steps to an ecology of mind*. New York: Ballantine.

Beer, S. (1986). *The brain of the firm*, 2nd ed. New York: Penguin Press.

Bélanger, M., Allaman, I., & Magistretti, P. J. (2011). Brain energy metabolism: Focus on astrocyte-neuron metabolic cooperation. *Cell Metabolism*, 14(6), 724–738.

Bessi, A., & Ferrara, E. (2016). Social bots distort the 2016 U.S. presidential election online discussion. *First Monday*, 21(11). http://firstmonday.org/ojs/index.php/fm/article/view/7090/5653

Bitran, H. (2016). Microsoft Health Bot Project. www.microsoft.com/en-us/research/project/health-bot/

Boshmaf, Y., Muslukhov, I., Beznosov, K., & Ripeanu, M. (2013). Design and analysis of a social botnet. *Computer Networks*, 57(2), 556–578. doi: 10.1016/j.comnet.2012.06.006

Borowitz, A. (2017). Man in hostage video forced to recite words not his own. *New Yorker*, August 14. www.newyorker.com/humor/borowitz-report/man-in-hostage-video-forced-to-recite-words-not-his-own

Bosker, B. (2016). The binge breaker. *The Atlantic*, November. www.theatlantic.com/magazine/archive/2016/11/the-binge-breaker/501122/

Centola, D. (2010). The spread of behavior in an online social network experiment. *Science*, 329(5996), 1194–1197.

Centola, D. (2011). An experimental study of homophily in the adoption of health behavior. *Science*, 334(6060), 1269–1272.

Centola, D., & Macy, M. (2007). Complex contagions and the weakness of long ties. *American Journal of Sociology*, 113(3), 702–734.

Colby, K., Watt, J., & Gilbert, J. (1966). A computer method of psychotherapy. *The Journal of Nervous and Mental Disease*, 142(2), 148–152.

Coldewey, D. (2012). Romney Twitter account gets upsurge in fake followers, but from where? *NBC News*, August 8. www.nbcnews.com

Condliffe, J. (2017). Will self-driving cars be better drivers if they can chat with each other? *MIT Technology Review*, April 24. www.technologyreview.com/s/604258/will-self-driving-cars-be-better-drivers-if-they-can-chat-with-each-other/

Coren, M. (2016). Virtual assistants spend much of their time fending off sexual harassment. *Quartz*, October 25. https://qz.com/818151/virtual-assistant-bots-like-siri-alexa-and-cortana-spend-much-of-their-time-fending-off-sexual-harassment/

Davidson, D. (2017). Facebook targets "insecure" young people. *The Australian*, May 1. www.theaustralian.com.au/business/media/digital/facebook-targets-insecure-young-people-to-sell-ads/news-story/a89949ad016eee7d7a61c3c30c909fa6

Davis, W. (2009). *The wayfinders: Why ancient wisdom matters in the modern world*. Toronto, ON: Anasi.

Dwyer, R., & Fraser, S. (2016). Addicting via hashtags: How is Twitter making addiction? *Contemporary Drug Problems*, 43(1), 79–97. doi: 10.1177/0091450916637468

Edwards, C., Edwards, A., Spence, P., & Ashleigh, K. (2014). Is that a bot running the social media feed? Testing the differences in perceptions of communication quality for a human agent and a bot agent on Twitter. *Computers in Human Behavior*, 33(2014), 372–376. doi: 10.1016/j.chb.2013.08.013

Elishar, A., Fire, M., Kagan, D., & Elovici, Y. (2012). Organizational intrusion: Organization mining using socialbots. *Social Informatics 2012 International Conference* (pp. 7–12). Piscataway, NJ: Institute of Electrical and Electronics Engineers (IEEE). doi: 10.1109/SocialInformatics.2012.39

Ferguson, N. (2017). The False Prophecy of Hyperconnection. Foreign Affairs, 96(5): 68–79. https://www.foreignaffairs.com/articles/2017-08-15/false-prophecy-hyperconnection.

Fessler, L. (2017). We tested bots like Siri and Alexa to see who would stand up to sexual harassment. *Quartz*, February 22. https://qz.com/911681/we-tested-apples-siri-amazon-echos-alexa-microsofts-cortana-and-googles-google-home-to-see-which-personal-assistant-bots-stand-up-for-themselves-in-the-face-of-sexual-harassment/

Fields, D. (2011). *The other brain*. New York: Simon & Schuster.

Finn, E. (2017). *What algorithms want*. Cambridge, MA: MIT Press.

Floridi, L. (2014a). Artificial agents and their moral nature. In P. Kroes & P. P. Verbeek (Eds.), *The moral status of technical artifacts* (pp. 185–212). Dordrecht, The Netherlands: Springer Science and Business Media. doi: 10.1007/978-94-007-7914-3-11

Floridi, L. (2014b). *The fourth revolution: How the infosphere is reshaping human reality*. Oxford: Oxford University Press.

Forelle, M., Howard, P., Monroy-Hernandez, A., & Savage, S. (2015). Political bots and the manipulation of public opinion in Venezuela. http://arxiv.org/abs/1507.07109

Freedom House. (2015). United States. In *Freedom on the net 2015: Privatizing censorship, eroding privacy* (pp. 872–894). www.freedomhouse.org/

Gehl, R. (2014). *Reverse engineering social media: Software, culture, and political economy in new media capitalism*. Philadelphia: Temple University Press.

Geiger, S. (2009). The social roles of bots and assisted editing programs. *Proceedings of the 5th International Symposium on Wikis and Open Collaboration*. New York: ACM.

Geiger, S. (2014). Bots, bespoke, code and the materiality of software platforms. *Information, Communication, and Society*, 17(3), 342–356.

Geiger, S. (2016). Bot-based collective blocklists in Twitter: The counterpublic moderations of harassment in a networked public space. *Information, Communication, and Society*, 19(6), 787–803.

Goethe, J. W. (1790[1970]). *The metamorphosis of plants*. Cambridge, MA: MIT Press.

Goethe, J. W. (1810[1970]). *Theory of colours*. Cambridge, MA: MIT Press.

Goffman, W., & Newill, V. (1964). Generalization of epidemic theory: An application to the transmission of ideas. *Nature*, 4955(204), 225–228.

Gorwa, R., & Guilbeault, D. (2017). Tinder nightmares: The promises and perils of political bots. *Wired*, July 7. www.wired.co.uk/article/tinder-political-bots-jeremy-corbyn-labour

Graham, T., & Ackland, R. (2017). Do socialbots dream of popping the filter bubble? The role of socialbots in promoting deliberative democracy in social media. In R. Gehl and M. Bakardjieva (Eds.), *Socialbots and their friends: Digital media and the automation of society*. New York: Routledge, 187–206.

Guilbeault, D. (2016). Growing bot security: An ecological view of bot agency. *The International Journal of Communication*, 10, 5003–5021. http://ijoc.org/index.php/ijoc/article/view/6135

Guilbeault, D., & Woolley, S. (2016). How Twitter bots are shaping the election. *The Atlantic*, November 1. www.theatlantic.com/technology/archive/2016/11/election-bots/506072/

Guyunn, J. (2017). Facebook can tell when users feel insecure. *USA Today*, May 1. www.usatoday.com/story/tech/news/2017/05/01/facebook-can-tell-when-teens-feel-insecure-advertiser-target/101158752/#

Han, X., Chen, M., Wang, F., Windrem, M., Wang, S., Shanz, S., … Nedergaard, M. (2013). Forebrain engraftment by human glial progenitor cells enhances synaptic plasticity and learning in adult mice. *Cell Stem Cell*, 12(3), 342–353.

Hannan, M., & Freeman, J. (1989). *Organizational ecology*. Cambridge, MA: Harvard University Press.

Hayles, K. (1999). *How we became posthuman: Virtual bodies in cybernetics, literature and informatics*. Chicago: The University of Chicago Press.

Hidalgo, C. (2015). *Why information grows: The evolution of order, from atoms to economies*. New York: Basic Books.

Hine, G., Onaolapo, J., De Cristofaro, E., Kourtellis, N., Leontiadis, I., Samaras, R., … Blackburn, J. (2017). Kek, cucks, and god emperor Trump: A measurement study of 4chan's politically incorrect forum and its effects on the web. *ArXiv*. https://arxiv.org/abs/1610.03452

Hwang, T., Pearce, I., & Nanis, M. (2012). Socialbots: Voices from the front. *Social Mediator*, March–April, pp. 38–45. doi: 10.1145/2090150.2091061

Just, M., Crigler, A., Metaxes, P., & Mustafaraj, E. (2012). "It's trending on Twitter": An analysis of the Twitter manipulations in the Massachussets 2010 special senate election. Paper presented at the *Annual Meeting of the American Political Science Association*. New Orleans, LA, August 30–September 2. https://papers.ssrn.com/sol3/papers.cfm?abstract_id=2108272

Keesey, I., Koerte, S., Khallaf, M., Retzke, T., Guillou, A., Grosse-Wilde, E., … Hansson, B. (2017). Pathogenic bacteria enhance dispersal through alteration of *Drosophila* social communication. *Nature Communications*, 8(265). www.nature.com/articles/s41467-017-00334-9

Kennedy, K. (2009). Textual machinery: Authorial agency and bot-written texts in Wikipedia. In M. Smith & B. Warnick (Eds.), *The responsibilities of rhetoric* (pp. 303–309). Long Grove, IL: Waveland.

King, G., Pan, J., & Roberts, M. (2016) How the Chinese government fabricates social media posts for strategic distraction, not engaged argument. *American Political Science Review*, 11(3), 484–501.

Krafft, P., Macy, M., & Pentland, A. (2016). Bots as virtual confederates: Design and ethics. *The 20th ACM conference on computer-supported cooperative work and social computing (CSCW)*. doi: 10.1145/2998181.2998354.

Krippendorff, K. (1986). *A dictionary of cybernetics.* Norfolk, VA: American Society for Cybernetics. http://repository.upenn.edu/asc_papers/224

Kuss, D., & Griffiths, M. (2011). Online social networking and addiction: A review of the psychological literature. *International Journal of Environmental Research and Public Health*, 8(9), 3528–3552. doi: 10.3390/ijerph8093528

Laorden, C., Galan-Garcia, P., Santos, I., Sanz, B., Hidalgo, J., & Bringas, P. (2013). Negobot: A conversational agent based on game theory for the detection of paedophile behaviour. *International joint conference CISIS'12-ICEUTE'12-SOCO'12 special sessions*. Berlin and Heidelberg: Springer.

LaRose, R., Kim, J., & Peng, W. (2011). Social networking: Addictive, compulsive, problematic, or just another media habit? In Z. Papacharissi (Ed.), *A networked self: Identity, community, and culture on social network sites* (pp. 59–82). New York: Routledge.

Le Guin, U. (1974). *The dispossessed*. New York: Harper Collins.

Leonard, A. (1998). *Bots: The origin of new species*. New York: Penguin.

Lucas, G., Gratch, J., King, A., & Morency, L. (2014). It's only a computer: Virtual humans increase willingness to disclose. *Computers in Human Behaviour*, 37, 94–100.

Malbon, J. (2013). Taking fake online consumer reviews seriously. *Journal of Consumer Policy*, 36(2), 139–157.

Martinez, A. (2016). *Chaos monkeys: Obscene fortune and random failure in Silicon Valley*. New York: Harper.

Maturana, H., & Varela, F. (1979). *Autopoiesis and cognition: The realization of the living*. Boston, MA: D. Reidel Publishing.

Maxwell, J. C. (1882). Are there real analogies in nature? In L. Campbell and W. Garnett (Eds.), *Life of James Clerk Maxwell* (pp. 235–244). London: Macmillan.

McLuhan, M. (1964). *Understanding media: The extensions of media*. New York: McGraw Hill.

McKelvey, F., & Dubois, E. (2017). Computational propaganda in Canada: The use of political bots. In S. Woolley and P. N. Howard (Eds.), *Working Paper* 2017.6. Oxford: Project on Computational Propaganda. http://comprop.oii.ox.ac.uk/

Messias, J., Schmidt, L., Oliveira, R., & Benevenuto, F. (2013). You followed my bot! Transforming robots into influential users on Twitter. *First Monday*, 18(7). http://firstmonday.org/ojs/index.php/fm/article/view/4217/3700#p4

Metaxas, P.T., & Mustafaraj, E. (2012). Social media and the elections. *Science*, 338(6106), 472–473. doi: 10.1126/science.1230456

Metzger, M., & Flanagin, A. (2013). Credibility and trust of information in online environments: The use of cognitive heuristics. *Journal of Pragmatics*, 59, 210–220. doi: 10.1016/j. pragma.2013.07.012

Miyamoto, A., Wake, H., Moorhouse, A., & Nabekura, J. (2013). Microglia and synapse interactions: Fine tuning neural circuits and candidate molecules. *Frontiers in Cellular Neuroscience*, 7(70). doi: 10.3389/fncel.2013.00070

Molteni, M. (2017). The chatbot therapist will see you now. *Wired*, June 7. www.wired. com/2017/06/facebook-messenger-woebot-chatbot-therapist/

Mønsted, B., Sapiezynski, P., Ferrara, E., & Lehmann, S. (2017). Evidence of complex contagion of information in social media: An experiment using Twitter bots. *PloS One*, 12(9), e0184148.

Morisano, D., Hirsh, J., Peterson, J., Pihl, R., & Shore, M. (2010). Personal goal setting, reflection, and elaboration improves academic performance in university students. *Journal of Applied Psychology*, 95, 255–264.

Munger, K. (2016). Tweetment effects on the tweeted: Experimentally reducing racist harassment. *Political Behavior*, 39, 629–649.

Naess, A. (1995). Self-realization: An ecological approach to being in the world. In G. Sessions (Ed.), *Deep ecology for the twenty-first century* (pp. 225–239). Boston, MA: Shambhala.

Neff, G., & D. Nafus. (2016). *Self-tracking*. Cambridge, MA: MIT Press.

Neff, G., & P. Nagy. (2016). Talking to bots: Symbiotic agency and the case of Tay. *International Journal of Communication*, 10, 4915–4931.

Newton, C. (2016). Speak, memory: When her best friend died, she rebuilt him using artificial intelligence. *Verge*. www.theverge.com/a/luka-artificial-intelligence-memorial-roman-mazurenko-bot

Pachter, J. (2003). The blood-brain barrier and its role in immune privilege in the central nervous system. *Journal of Neuropathology & Experimental Neurology*, 62(6), 593–604.

Page, S. (2008). *The difference: How the power of diversity creates better groups, firms, schools, and societies*. Princeton: Princeton University Press.

Papacharissi, Z. (2009). The virtual geographies of social networks: A comparative analysis of Facebook, LinkedIn, and ASmallWorld. *New Media & Society*, 11(1–2), 199–220. doi: 10.1177/1461444808099577

Papacharissi, Z. (2012). Without you, I'm nothing: Performances of the self on Twitter. *International Journal of Communication*, 6. http://ijoc.org/index.php/ijoc/article/view/1484

Pareene, A. (2016). How we fooled Donald Trump into Retweeting Benito Mussolini. *Gawker*, February 28. http://gawker.com/how-we-fooled-donald-trump-into-retweeting-benito-musso-1761795039

Parisi, L. (2013). *Contagious architecture: Computation, aesthetics, and space*. Cambridge, MA: MIT Press.

Pask, G. (1969). The architectural relevance of cybernetics. *Architectural Design*, 39 (September), 494–496.

Payrs, B., Cote, A., Gallo, V., De Koninck, P., & Sik, A. (2010). Intercellular calcium signaling between astrocytes and oligodendrocytes via gap junctions in culture. *Neuroscience*, 167(4), 1032–1043.

Pennebaker, J., W. & Stone, L. D. (2004). Translating traumatic experiences into language: Implications for child abuse and long-term health. In L. J. Koenig, L. S.

Doll, A. O'Leary, & W. Pequegnat (Eds.), *From child sexual abuse to adult sexual risk: Trauma, revictimization, and intervention* (pp. 201–216). Washington, DC: American Psychological Association.

Pickering, A. (2010). *The cybernetic brain: Sketches of another future.* Chicago: Chicago University Press.

Ratkiewicz, J., Conover, M., Meiss, M., Goncalves, B., Flammini, A., & Menczer, F. (2011a). Detecting and tracking political abuse in social media. *Proceedings of the fifth international AAAI conference on weblogs and social media.* www.aaai.org/ocs/index.php/ICWSM/ICWSM11/paper/view/2850

Ratkiewicz, J., Conover, M., Meiss, M., Goncalves, B., Patil, S., Flammini, A., & Menczer, F. (2011b). Truthy: Mapping the spread of astroturf in microblog streams. *WWW '11: Proceedings of the 20th international conference companion on world wide web* (pp. 249–252). doi: 10.1145/1963192.1963301

Rodgers, D. (2012). *Age of Fracture.* Cambridge, MA: Harvard University Press.

Salisbury, John of (Bishop of Chartres). (1159/1979). *Policraticus: The statesman's book.* New York: F. Ungar.

Savage, S., Monroy-Hernandez, A., & Hollerer, T. (2016). Botivist: Calling volunteers to action using online bots. *Proceedings of the 19th ACM conference on computer-supported cooperative work & social computing (CSCW)* (pp. 813–822). New York: ACM.

Schneier, B. (2014.) *Data and Goliath: The hidden battles to collect your data and control your world.* New York: W. W. Norton.

Shirado, H., & Christakis, N. (2017). Locally noisy autonomous agents improve global human coordination in network experiments. *Nature, 545,* 370–374.

Smyth, J. M. (1998). Written emotional expression: Effect sizes, outcome types, and moderating variables. *Journal of Consulting and Clinical Psychology, 66*(1), 174–184.

Spencer, H. (1864). *Principles of biology.* London: Williams and Norgate.

Spencer, H. (1896). *The study of sociology.* New York: D. Appleton.

Stryker, C. (2011). *Epic win for Anonymous: How 4chan's army conquered the web.* New York: Overlook Duckworth

Subrahmanian, V. S., Azaria, A., Durst, S., Kagan, V., Galstyan, A., Lerman, K., … Hwang, T. (2016). The DARPA Twitter bot challenge. *Computer, 49*(6), 38–46.. doi: 10.1109/MC.2016.183

Tsvetkova, M., Garcia-Gavilanes, R., Floridi, L., & Yasseri, T. (2017). Even good bots fight: The case of Wikipedia. *PLoS ONE, 12*(2): e0171774. doi: 10.1371/journal.pone.0171774

Turkle, S. (1995). *Identity on the screen.* New York: Simon & Schuster.

Turkle, S. (2011). *Alone together.* New York: Basic Books.

Turow, J. (2011). *The daily you.* New Haven, CT: Yale University Press.

Turow, J. (2017). *The aisles have eyes: How retailers track your shopping, strip your privacy, and define your power.* New Haven, CT: Yale University Press.

van Dijck, J. (2013). "You have one identity": Performing the self on Facebook and LinkedIn. *Media, Culture & Society, 35*(2), 199–215. doi: 10.1177/0163443712468605

Verkhratsky, A., & Nedergaard, M. (2016). The homeostatic astroglia emerges from evolutionary specialization of neural cells. *Philosophical Transactions of the Royal Society, 371* (1700). doi: 10.1098/rstb.2015.0428

Vincent, J. (2016). Baidu launches medical chatbot to help Chinese doctors diagnose patients. *The Verge.* www.theverge.com/2016/10/11/13240434/baidu-medical-chatbot-china-melody

Wade, J., McDaid, L., Harkin, J., Crunelli, V., & Kelso, J. (2011). Bidirectional coupling between astrocytes and neurons mediates learning and dynamic coordination in the brain: A multiple modeling approach. *PLoS One*, 6(12), e29445.

Wallace, D. (1996). *Infinite jest*. New York: Little Brown.

Weizenbaum, J. (1966). ELIZA—a computer program for the study of natural language communication between man and machine. *Communications of the ACM*, 9, 36–45.

Wiener, N. (1950). *The human use of human beings: Cybernetics and society*. New York: Avon Books.

Wininger, S. (2016). The secret behind Lemonade's instant insurance. *Lemonade*, November 23. https://medium.com/@shai_wininger/the-secret-behind-lemonades-instant-insurance-3129537d661#.89a6k6lkp

Wong, S. (2016). Google translate AI invents its own language to translate with. *New Scientist*, November 30. www.newscientist.com/article/2114748-google-translate-ai-invents-its-own-language-to-translate-with/

Woolley, S. C. (2016). Automating power: Social bot interference in global politics. *First Monday*, 21(4). http://firstmonday.org/ojs/index.php/fm/article/view/6161/5300

Woolley, S., & Guilbeault, D. (2017). Computational propaganda in the United States of America: Manufacturing consensus online. In S. Woolley and P. N. Howard (Eds.), *Working Paper 2017.5*. Oxford: Project on Computational Propaganda. comprop.oii. ox.ac.uk. http://comprop.oii.ox.ac.uk/

Woolley, S., & Howard, P. (2017). Computational propaganda worldwide: Executive summary. In S. Woolley and P. N. Howard (Eds.), *Working Paper 2017.1*. Oxford: Project on Computational Propaganda. comprop.oii.ox.ac.uk. http://comprop.oii.ox.ac.uk/

Xie, Y., Yu, F., Ke, Q., Abadi, M., Gillum, E., Vitaldevaria, K., … Morley, M. Z. (2012). Innocent by association: Early recognition of legitimate users. *Proceedings of the 2012 ACM conference on computer and communications* (pp. 353–364). New York: ACM. doi: 10.1.1.299.129

Zittrain, J. (2008). *The future of the internet and how to stop it*. New Haven, CT: Yale University Press.

13

AI, THE PERSONA, AND RIGHTS

Tamara Shepherd

The notion of rights in digital culture is a fraught one. Frameworks for inalienable human rights, themselves critiqued on the grounds of abstraction, Eurocentrism, and gender bias, have been subject to intense debate when transposed to networked spaces. The 2017 RightsCon in Belgium, for instance, featured talks on diverse aspects of internet rights, from regulatory jurisdiction to environmental impacts of technology to hate speech online to trade agreements. Germane to all of these themes, one conclusion that has emerged over the last 20-odd years of the internet's popularization has been a sense that digital culture is in some ways an inherently commercial construct, where ideals of democratization are merely rhetorical tools that shore up consolidated corporate power (e.g. Dean, 2005; Gillespie, 2010; Zuboff, 2015).

Given the overwhelmingly commercial nature of what most people experience as digital culture, and the location of that brand of commercialism predominantly in the United States (Jin, 2013), how might constitutional rights to personal integrity and freedom be considered? One framework for reinstating the legal viability of the person in digital culture is William McGeveran's (2009) formulation of "persona rights." McGeveran sees much of digital culture as defined by the boundaries of social platforms supported by advertising that commodifies users' endorsements; in other words, a "like" economy (e.g. Gerlitz & Helmond, 2013). Since user data (privacy) and creative content (intellectual property) is appropriated by social platforms, McGeveran argues, an individual should have recourse to their status as legal persona to exercise monopoly control over their own image. Persona rights thus offer a novel legal framework for bolstering the integrity of the person within commercial spaces for digital culture.

Of course, persona rights reflects a liberal context and can thus be critiqued on the grounds that it re-entrenches neoliberal individuality as the basis for rights, forgoing socialist, feminist, and Indigenous versions of collective rights. However, as promising as such approaches may be for resisting the encroachment of neoliberal rationality, the current configuration of digital culture is one controlled by a handful of consolidated techno elites. This is especially the case in contemporary advancements in artificial intelligence (AI): the development of machine learning based on tightly controlled proprietary algorithms, growing in complexity from recommender systems (e.g. Netflix) to deep learning (e.g. Google). By creating algorithms with the capacity to optimize prediction, machine-learning strategies are being used in diverse contexts from medical diagnosis to autonomous vehicles to weapons systems. Paying particular attention to the development of artificial general intelligence (AGI)—the capacity for computational machines to solve a variety of complex problems across domains by controlling themselves autonomously and even self-consciously (Pennachin & Goertzel, 2007)—this chapter examines how machine-learning algorithms add fundamental complications to the already fraught terrain of digital rights. By framing such rights in terms of persona rights, I hope to add to a growing suite of means to reassert the human as a rights-bearing agent within posthumanism.

The Posthuman Conceit of AI

Thinking machines have been framed as "posthuman persons" (Hamilton, 2009, p. 143), in that their interactions with human interlocutors form the basis for our perception of their intelligence. AI systems both learn from humans and take their inspiration from the human nervous system, materializing Donna Haraway's cyborg in a way that suggests a displacement of transformative agency onto machines. In this posthuman context, developments in AI exert the same pressure on the idea of the human as "special" that advances in our understandings of animal cognition do, which links the idea of posthuman rights to those of animal rights and positions humans as stewards of these other intelligences (Haraway, 1991, 2008; Hayles, 1999).

Posthumanism has been a salient concept within internet studies more generally, where networked communication rests on the unit of the individual user-person, whose subjectivity is configured through a "liberal strategy of individual self-worth" (Balakrishnan, 2016, p. 112). As social platforms have come to dominate most users' experiences of the internet, personalization has become the hallmark of popular platforms driven by data-based advertising models. This intense personalization rests on the AI manifested in machine-learning algorithms that benefit from huge amounts of data being inputted by users of platforms like Netflix, Facebook, Google, and Amazon. Developed by large technology companies, such machine-learning algorithms tend to be proprietary and non-transparent. The lack of transparency makes sense given the trade secrets that characterize technological development, but has significant implications when these algorithms move toward deep learning.

Deep learning is a subset of machine learning where the algorithm itself—rather than the programmer—defines features to be analyzed and how the outcomes should be optimized by modeling how the human brain works. Inputs to the algorithm are analyzed and processed. If the processing results in a correct outcome, those analytical pathways are reinforced; if the outcome is incorrect, the pathways are reconfigured (Sample, 2017). This sort of independent adjustment through learning is called backpropagation and conceptually resembles neural plasticity in human brains. Google has different divisions currently working on deep learning, including Google Brain and Google DeepMind. These divisions are working on backpropagation models but also developing an alternative that uses even less computing power, which is called the evolutionary strategies approach (Domingos, 2015). The evolutionary strategies model even more closely resembles how the brain works, where information only travels forward—not backward—across neuronal synapses (Lake et al., in press, p. 36). And the name for this process, "evolutionary," more closely aligns AI rhetorically with the brain by using the language of a particularly human view of evolutionary adaptation: "Humans are the species that adapts the world to itself instead of adapting itself to the world. Machine learning is the newest chapter in this million-year saga" (Domingos, 2015, p. 3). Deep learning represents a future of machines both adapting themselves to the world and in turn adapting the world to themselves, no longer subject to supervised learning under the guidance of programmers but engaged in reinforcement learning and autonomous decision making (Stone et al., 2016, p. 15).

When considering advancements in AI systems through the prism of posthumanism, perhaps the most tantalizing possibility is that of artificial general intelligence (AGI). Today's deep learning AIs are trained to learn specific tasks such as object or speech recognition (Stone et al., 2016; Weston et al., 2014). AGI presents a general purpose alternative that can handle a diverse set of tasks by drawing on a more humanlike flexible memory (Pennachin & Goertzel, 2007). Such flexibility is made possible by mimicking the functional structure of the brain through neuromorphic computing. Neuromorphic computing replaces traditional computational models that separate inputs/outputs, instruction-processing, and memory, instead integrating these modules in a way similar to the brain (Stone et al., 2016, p. 9). Deep learning researchers foresee AGI arising from such attempts to get closer and closer to brain-like processing, where future neural networks are projected to be "endowed with intuitive physics, theory of mind, causal reasoning, and other capacities," including the capacity to "effectively search for and discover new mental models or intuitive theories" that can be used as the basis for subsequent learning (Lake et al., in press, p. 3).

In order to reach such a posthuman ideal, though, AGI must overcome a series of obstacles in its evolutionary progression. One significant barrier is computing power: an average brain contains about 100 billion neurons, each connected to up to 10,000 other neurons through a total of between 100 trillion and

1000 trillion synapses. To match that in computing power will require leaps in processing technology, perhaps only achievable through parallel or quantum computing. Nonetheless, it seems that AGI is somewhere on the horizon. And even though machines think in ways that are distinct from human thinking, deep learning still relies on homophily to the brain in terms of structural inspiration from neuroscience (Le et al., 2012), as well as psychological benchmarks that rest on comparisons with human intelligence (Kahn et al., 2007). As N. Katherine Hayles (1999) contends about AI, its trick lies in the tests (including famously, the Turing Test) used to assess intelligence. If human intelligence is the benchmark for determining the effectiveness of AGI, we might miss the distinctly posthuman capacities that come from alternative models of learning (Stone et al., 2016, p. 13). Already, AI outstrips human capacities in tasks such as instantaneous prediction, and this is where posthumanism suggests the need for a politics. Given the development and deployment of AI within technology companies emblematic of late capitalism, what is needed is an extension of Winner's (1977) sense of technology as having a politics to account for AI's potential reshaping of democratic rights in a technocractic context (Damnjanović, 2015).

Rights and Commerce

It is crucial to consider the posthuman context when adapting conceptions of rights, typically framed as liberal, humanist rights, to the specific situation of AI. Human rights is a relatively new concept, with that phrase only in common usage since the 1940s and the adoption of the United Nations' Universal Declaration of Human Rights (UDHR). Yet the idea of inalienable rights of the person extends much further back in Western thought (from Ancient Greek societies through the Middle Ages), and concepts such as "natural rights" can be seen to come into their own as the cornerstone of Enlightenment philosophy. For example, in his *Second Treatise Concerning Civil Government* (1689), John Locke argued for men's natural right to life, liberty, and property. Such arguments furnished state-level rights protections, particularly in contemporary revolutionary contexts, as evidenced in the English Bill of Rights (1689), the American Declaration of Independence (1776), and the French Déclaration des droits de l'homme et du citoyen (1789).

While all of these implementations of the idea of natural rights differ according to their specific political contexts, they share a notion that people should be naturally endowed with certain basic rights, protections or freedoms that support respect for the person's agency as a human being. These kinds of rights are framed by current human rights discourse as moral rights—rights attributable to a person simply because they exist as a human being—and should thus be distinguished from legal rights, or rights upheld by courts in a specific jurisdiction. Human rights treatises attempt to recognize moral rights as legal rights, although this is complicated by diverse understandings of both, subject to historical circumstance. For example, Kantian moral philosophy posits morality as fundamentally rational and moral responsibility as the basis for rights. Each person's self-governing reason to act dutifully (in accordance with the categorical imperative), in Kant's view, provided a deontological version of rights whereby that person could be viewed as possessing equal worth and deserving of equal respect as another. Yet the universalizability of Kantian moral philosophy, while it accords with ideas about universal human rights, exists in abstraction more so than in lived reality. A more situated historical approach to rights as ethical rather than moral positions human rights more directly within contemporary political contexts by seeking to define what exactly is meant by rights under the current "human rights regime" (e.g. Rawls, 2001).

The current regime of the UDHR reflects something of a continuity between the Enlightenment moment for natural rights within emergent capitalism and the late capitalist context of the twentieth century. In both eras, ideas of personhood coincided with models for extracting surplus value from the person in the form of labor power, in Marxian terms. Legally recognized as "persona," personhood has come to take on a commodity form, where "the persona confirms the labor and authoring capacity of the individual person" (Hamilton, 2009, p. 190). What changes with digital culture is the transition to surveillance capitalism (Zuboff, 2015), which warps the values of individualism central to liberal and neoliberal versions of the persona as a rights-bearing agent by segmenting individuals into even smaller units—"dividuals" (Deleuze, 1992; Terranova, 2004)—through data collection regimes that feed into AI systems on networked platforms.

While humans get thus reconfigured as dividuals through their contractual relationships with technology companies that run social platforms, the evolution from AI to AGI suggests a new formulation of posthuman rights. Anthropomorphizing machines has consequences for legal frameworks around agency (Hamilton, 2009, p. 175). For instance, a 2016 proposal to the European Parliament recommends granting personhood status to thinking machines according to the EU Charter of Fundamental Rights (Delvaux, 2016). Largely, this appeal to the persona of AI rests on concerns about liability, where machines might be held legally responsible for their own actions. The proposal goes on to suggest that the autonomy of advanced AI systems "raises the question of their nature in the light of the existing legal categories—of whether they should be regarded as natural persons, legal persons, animals or objects—or whether a new category should be created" (Delvaux, 2016, p. 5). The concern here is the legal status of AI in relation to existing property regimes, including the way in which machine learning relies on the collection of "personal data as a 'currency' with which services can be 'bought,'" challenging other EU regulations around data protection (Delvaux, 2016, p. 8). This proposal is remarkable for both noting the confounding categorization of AI in relation to rights while also positioning contemporary legal rights frameworks squarely within surveillance capitalism's version of data as licensed property and currency.

Persona Rights in AI

Private law of contract, manifested in social platforms' Terms of Service agreements, exerts significant pressure on contemporary versions of rights under surveillance capitalism. The whole idea of data as currency gets legally recognized through Terms of Service that most internet users do not even read (Klang, 2004), but that dictate the contractual relationship between the persona and the platform. According to the common stipulations of Terms of Service agreements, platforms commodify the content and personal information that users provide. As Lawrence Lessig (2006) explains, private law of contract thus displaces government regulation on the internet, according to the profit motive (pp. 185–187). While users can exert some influence over the fairness of these contracts through traditional channels like the court system, or new technological channels like the development of alternative internet architectures, large technology companies still dominate. Further, while Terms of Service agreements vary by website and host country, many of the most popular sites globally are based in the United States (Jin, 2013). As such, state-sponsored policy around users' rights needs to consider not only the commercialization of persona, but also the geopolitical imbalance of such control.

More broadly, a Western bias underpins the contractual relationship between platform and persona through the form of the Terms of Service agreement itself. As Carole Pateman (1988) argues, the way that one's identity or persona can generate proprietary interest under capital emerges from the libertarian legacy of self-ownership. The idea of owning oneself translates into alienable rights to "property in the person," which subjects the person, as private property, to a framework she calls "contractarianism." Contractarianism contains an inherent paradox for Pateman, given that contracts binding the person—such as employment contracts—alienate the person-as-laborer's right of self-government while simultaneously demanding labor as a humanistic practice, a practice marked by self-government and autonomy (Pateman, 2002, p. 47). A parallel paradox is apparent in social media platforms' Terms of Service: platforms "retain licensing rights over user content and expression, amid celebrations of user agency and the democratization of cultural production" (Shepherd, 2012, p. 107). This disjuncture might be taken even further in the context of machine learning, where the contract users enter into suggests full access to their private data and intellectual property as resources on which to train platforms to better personalize and meet user needs, or indeed, solve complex global problems (Zuckerberg, 2017).

Under surveillance capital, however, the idea of the individual persona is challenged by the way Terms of Service contracts facilitate the splitting up of individuals into dividuals—data derivatives that can be aggregated and reconfigured according to the pattern-generating logic of big data (Amoore, 2011). In this context, individual rights within contractarianism become even more tenuous. For example, American legal frameworks for intellectual property have only strengthened corporate claims to digital culture, as in the Digital Millennium Copyright Act (1998), and Federal Communications Commission privacy laws were repealed in early 2017 by Congress and the Trump Administration. Given these kinds of legal challenges to the integrity of individual privacy and intellectual property, which only stand to increase in the context of AI, it seems important to attempt to reinstate the legal salience of the person—despite its shortcomings—as a means to reassert even basic versions of liberal rights.

As a legal concept, the term "persona rights" refers to an individual's control over commercial appropriation of their identity in social media platforms. McGeveran (2009) explains how, within the context of behavioral advertising practices, persona rights law "transcends the narrower focus of other paradigms on protecting information privacy or preventing misleading advertising" (p. 1154). Persona rights open up more familiar understandings of online privacy or intellectual property by focusing on the practice of endorsement in social media platforms. Platforms like Facebook leverage their users' identities within algorithmic advertising models that deploy the "like" button (Gerlitz & Helmond, 2013), attempting to predict through machine learning what kinds of content or products will lead to greater user endorsement. For McGeveran, similar to celebrity endorsement, user endorsement might be subject to "two related but distinct legal claims: the tort of appropriation and the right of publicity" in US law (McGeveran, 2009, p. 1149). By protecting against unauthorized commercial uses of one's identity (the tort of appropriation) and maintaining monopoly control over one's own image (the right of publicity), these two legal instruments serve the function of recognizing the integrity and dignity of personal identity, in light of commercial exploitation, as sanctioned by the state. Persona rights law thus rests on the premise that individual users should have recourse to legal and regulatory protection of their rights to control their personal identities online, even as they get reconfigured into data derivatives used as inputs for machine learning.

While the persona rights framework has its limitations, including maintaining neoliberal individualism, proving dignitary harm (McGeveran, 2009), conflating privacy with property (Samuelson, 2000), and legal jurisdiction of specific tort law (Hamilton, 2009), the concept of persona rights offers a reinstatement of personal integrity in a posthuman context where AGI threatens to displace the concept of identity even further under surveillance capitalism. So far, due to the lack of robust state protections, suggestions for how to tackle the problem of endorsement underlying commercial AI have largely come from within the industry as it seeks to avoid alienating user-consumers. For example, Google has proposed a privacy solution that combines federated learning with differential privacy. Federated learning entails training AI systems with data that is aggregated rather than connected to an individual's identity (McMahan et al., 2017). This would be done through encrypted updates that prevent Google from seeing sensitive personal information, however, in practice, it would be possible to reverse engineer those updates to reveal the training data. Differential privacy is a strategy devised by

Apple that applies complex mathematical techniques to render sensitive information unidentifiable. Yet the drawback of this approach is that it would also entail users sending more data to Apple than ever before (Greenberg, 2016). Apparently, the impetus of persona rights to re-focus legal protection on the person will remain crucial as AI training systems develop.

From another perspective, moreover, persona rights reinstate not only the legal salience of the person but also the specificity of the human person within rights frameworks that are encroached upon by non-human or posthuman persons. As Peter Kahn and his co-authors suggest in their article "What is a human?" (2007), AI stretches the limits of humanity by being essentially determined by its ability to meet human psychological benchmarks. AI's success, they argue, is reliant on mimicking humanlike capacities through "categories of interaction that capture conceptually fundamental aspects of human life" (Kahn et al., 2007, p. 366). One of these is creativity or the capacity of AI to generate what might be recognized as intellectual property. As the authors note, even if AI cannot be said to develop consciousness, the viability of any AGI would be fundamentally predicated on creativity in approaching a novel task using past learning unrelated to that specific task (Kahn et al., 2007, p. 378). In terms of privacy, AI represents accelerated, immanent, and intimate data collection that stands to undermine the idea of contextual integrity altogether (Kahn et al., 2007, p. 373; Nissenbaum, 2004). Finally, perhaps the ultimate benchmark for AI is autonomy, which speaks to the heart of posthumanism by acknowledging that autonomy for both humans and machines is subject to debate about the degree to which decision making is conditioned by internal and external factors. In liberal formulations of rights, humans are invested with autonomy and by extension moral responsibilities that accompany basic rights of dignity and integrity (e.g. Dworkin, 1978). For Kahn and colleagues (2007), autonomy, or at least autonomous behavior, as a benchmark for AI thus raises the issue of rights beyond just persona rights of privacy and intellectual property. While the authors concur that such benchmarks for determining the success of AI are not meant as ontological categories but rather functional, psychological ones (Kahn et al., 2007, p. 365), the benchmarks nonetheless illuminate the constructedness of rights frameworks alongside the constructedness of humanlike versions of privacy and intellectual property in AI.

Regulating AI

The question so far addressed in this chapter about the status of persona rights given the development of AI and AGI leads naturally to a consideration of legal protections for liberal rights. Especially around privacy and intellectual property, advancements in machine learning demand new frameworks for understanding the regulatory implications of applications already in use, such as assistant AI systems (Apple's Siri, Amazon Echo, Google Home) that collect ambient voice data and computational creativity systems like Adobe Sensei that automates design.

Add to these the potential leaps in AGI—depicted evocatively, for example, in Louisa Hall's recent novel *Speak* (2015), where sentient robots become friends for withdrawn children—and it seems that regulation has a difficult task ahead to keep pace with the diverse implications of technological developments.

Perhaps the greatest challenge in devising new regulatory frameworks is the opacity of the technology. AI systems are typically inscrutable in terms of how certain inputs result in outputs, in what has been termed the explainability problem (Heaven, 2013). This is apparent in the European Parliament proposal mentioned above, where the status of AI is itself unclear (Delvaux, 2016). So, even though the US government has made attempts to leverage technical expertise (Stone et al., 2016, p. 10)—for example, in the task force advising on the May 2016 report of the National Science and Technology Council, "Preparing for the Future of Artificial Intelligence"—there are limits to regulating something that cannot be fully understood even by developers. Nonetheless, the report suggests seven overarching mandates in AI policy, including that AI should be used for public good, bias must be eliminated from data, and global partnerships are necessary to ensure transparency.

The co-articulation of public good, bias, and transparency in this regulatory proposal suggests a link back to liberal conceptions of rights where individuals might expect fair and equal treatment under the law. And yet eliminating bias from AI is as tricky a proposition as eliminating bias from humans. Algorithms, it has been argued, are inherently ideological; they represent a "machinic subjectivity" that interpellates us through our data (Jones, 2014, p. 251). For instance, much has been said about the way AI systems enact a discriminatory reinforcement of existing social inequalities (Crawford, 2016; Leurs & Shepherd, 2016). Often these critiques rest on a supervised learning model where AI is fed data labeled by people and AI systems rest on invisible human labor (Gray & Suri, 2017). In reinforcement learning, however, unlabeled data is processed by algorithms and so it is not human but "algorithmic responsibility" that underlies discrimination (Banavar, 2016).

In order to police algorithmic responsibility without the need for restrictive government intervention, a number of industry-led initiatives have been proposed. This sort of self-regulation is seen to be potentially more effective than state policy due to non-specialist regulators' difficulty in keeping up with the technology (Banavar, 2016; Stone et al., 2016). For example, the Elon Musk-backed OpenAI project, announced in December 2015, focuses on developing AI for public good. The Partnership on AI, a collaborative effort involving Google, Facebook, Amazon, IBM, and Microsoft, was launched in September 2016. The founders of LinkedIn and eBay set up the Ethics and Governance of Artificial Intelligence Fund in January 2017 to support academic research on the safety of AI systems. These initiatives share a preoccupation with transparency around AI, matching calls from regulators and academics for third-party audits of algorithmic bias (Wachter et al., 2017). A "right to explanation" may help to ensure that industry-led initiatives live up to their promises for fair AI (Wachter et al., 2017), by providing clear explanations of what is behind the black box of algorithmic decision making so that people, even non-experts, can audit the process (Banavar, 2016, p. 4; Gray & Suri, 2017; Pasquale, 2015). This suggestion could fit in with the social systems analysis that Kate Crawford and Ryan Calo (2016) support for uncovering the social, political, and cultural values embedded in AI.

While increased transparency is an important and laudable goal for regulators, transparency alone is not sufficient to address the larger regulatory implications of AI and AGI. Returning to algorithms' explainability problem, the posthuman conceit of AI suggests that these systems work like the human brain (which itself is something that neuroscientists do not yet fully understand). Yet AI does not work like the brain, even though it takes inspiration from the brain. Instead, developments in deep learning come from computational rather than neuronal processes, and moreover, these processes often have unpredictable outcomes. As legal scholar Ian Kerr (2017) has argued, if AI developers cannot fulfill their "duty to explain," then unexplainable AI should be prohibited unless individuals can take recourse in "a right not to be subject to a decision that is based solely on automated processing." This proposal pivots on explainability as the issue that underlies transparency and represents a relatively interventionist stance on AI development as sociotechnical.

Such intervention may be necessary, despite attempts at industry self-regulation, given other, more long-standing problems with calls for transparency in the tech sector. To take transparency reports of the major technology companies as a prime example, the self-reporting done by industry players offers "a very particular kind of reporting, which may cater to demands for openness and disclosure about government surveillance and censorship, but provide a very specific response in a preformatted and selective shape" (Flyverbom, 2016, p. 8). Given that these same tech companies are behind recent advancements in AI, where machine learning is largely geared toward functions that can be monetized, this critique of transparency gains additional salience. Especially in the context of AGI, it is not difficult to imagine how it will be under attendant capitalist pressures to generate surplus value, continuing the imperatives of social media platforms to harness general intellect and flexible labor (Cohen, 2011; Jarrett, 2015). As the current limitations in processing power give way, machine learning will represent the "next frontier" for imperialist ambitions of tech companies to accelerate commodification. Perhaps the strongest limitations in this case will be philosophical. To return to the liberal political theory underpinning the idea of rights, the line between human and non-human or posthuman remains deeply invested by our own beliefs in the primacy of consciousness. Perhaps it is on these grounds that the persona can be recuperated along with its rights to self-determination.

Conclusion: Individual vs. Collective Rights

This chapter has deployed the persona rights framework to help consider the double challenge of AI to liberal rights: 1) that AI systems appropriate individuals' private data and intellectual property through machine learning algorithms; and 2) that the imminent development of AGI challenges the whole concept of persona through its posthuman conceit. Given the late capitalist context in which deep learning systems are emerging, where successful monetization of the technology rests on huge resources in data, persona rights offer something of a tactical response to maintaining human dignity and sovereignty. Through control over the uses of one's endorsement—encompassing personal data and creative output—persona rights bolster individual entitlement to control and use one's own persona (McGeveran, 2009, p. 1132).

Yet persona rights maintain a relatively liberal version of the individual as a coherent core over which sovereign control can be exercised—going against philosophies of technology that see the self as a social construct (Austin, 2010). Indeed, if the notion of a discrete self or persona might be challenged on philosophical grounds, it is also being undermined within digital culture as a society of control where dividual better describes the units through which people are made useful to machine learning as data derivatives (Amoore, 2011; Deleuze, 1992). The ways in which individual sovereignty fails as a precondition for rights suggests two paths forward: one in which the liberal individual is reconstituted—for example, through persona rights law—and the other in which a wholly new paradigm for collective rights might be articulated.

The loss of the sovereign individual thus merits thinking more strategically about resisting the larger configuration of neoliberal individualism within global capital. One place to start this thinking is through a reconfiguration of the idea of private property. Letting go of the notion of control over the self as property opens up onto alternative frameworks of the commons, as suggested by free culture advocates (e.g. Boyle, 2008; Lessig, 2006), feminist political philosophers (e.g. Pateman, 1988; Weinberg, 2017), and Indigenous studies scholars (e.g. Stabinsky & Brush, 2007). The questions to ask from this vantage point would be how collective rights might help maintain human dignity in an AI era: How can the colonization of general intelligence by tech companies be prevented? It seems particularly urgent to articulate posthuman rights before human and machine brains eventually merge in the envisioned singularity according to the efforts of technologists like Elon Musk. Given this urgency, what may look like anthropomorphism in the neural metaphors for AI, where machines are assigned subjectivity, can in fact portend a politics (Jones, 2014): the posthuman may be the privileged route into the postcapitalist.

References

Amoore, L. (2011). Data derivatives: On the emergence of a security risk calculus for our times. *Theory, Culture & Society*, 28(6), 24–43.

Austin, L. M. (2010). Control yourself, or at least your core self. *Bulletin of Science, Technology & Society*, 30(1), 26–29.

Balakrishnan, S. (2016). Historicizing hypertext and Web 2.0: Access, governmentality and cyborgs. *Journal of Creative Communications*, 11(2), 102–118.

Banavar, G. (2016). Learning to trust artificial intelligence systems: Accountability, compliance and ethics in the age of smart machines. White Paper prepared for IBM Corporation. www.research.ibm.com/software/IBMResearch/multimedia/AIEthics_Whitepaper.pdf

Boyle, J. (2008). *The public domain: Enclosing the commons of the mind*. New Haven, CT: Yale University Press.

Cohen, N. S. (2011). The valorization of surveillance: Towards a political economy of Facebook. *Democratic Communiqué*, 22(1), 5.

Crawford, K. (2016). Artificial intelligence's white guy problem. *New York Times*, 25 June. www.nytimes.com/2016/06/26/opinion/sunday/artificial-intelligences-white-guy-problem.html?_r=0

Crawford, K., & Calo, R. (2016). There is a blind spot in AI research. *Nature*, 538, 311–313.

Damnjanović, I. (2015). Polity without politics? Artificial intelligence versus democracy: Lessons from Neal Asher's Polity Universe. *Bulletin of Science, Technology & Society*, 35(3–4), 76–83.

Dean, J. (2005). Communicative capitalism: Circulation and the foreclosure of politics. *Cultural Politics*, 1(1), 51–74.

Deleuze, G. (1992). Postscript on the societies of control. *October*, 59, 3–7.

Delvaux, M. (2016). Draft report with recommendations to the Commission on Civil Law Rules on Robotics. European Parliament Committee on Legal Affairs, 31 May. www.europarl.europa.eu/committees/en/juri/draft-reports.html

Domingos, P. (2015). *The master algorithm: How the quest for the ultimate learning machine will remake our world*. New York: Basic Books.

Dworkin, R. (1978). *Taking rights seriously*. Cambridge, MA: Harvard University Press.

Flyverbom, M. (2016). Disclosing and concealing: Internet governance, information control and the management of visibility. *Internet Policy Review*, 5(3), 1–15.

Gerlitz, C., & Helmond, A. (2013). The like economy: Social buttons and the data-intensive web. *New Media & Society*, 15(8), 1348–1365.

Gillespie, T. (2010). The politics of "platforms." *New Media & Society*, 12(3), 347–364.

Gray, M. L., & Suri, S. (2017). The humans working behind the AI curtain. *Harvard Business Review*, 9 January. https://hbr.org/2017/01/the-humans-working-behind-the-ai-curtain

Greenberg, A. (2016). Apple's "differential privacy" is about collecting your data—but not *your* data. *Wired*, 13 June. www.wired.com/2016/06/apples-differential-privacy-collecting-data/

Hall, L. (2015). Speak: A Novel. New York: Ecco.

Hamilton, S. (2009). *Impersonations: Troubling the person in law and culture*. Toronto: University of Toronto Press.

Haraway, D. J. (2008). *When species meet*. Minneapolis: University of Minnesota Press.

Haraway, D. J. (1991). *Simians, cyborgs and women: The reinvention of nature*. New York: Routledge.

Hayles, N. K. (1999). *How we became posthuman: Virtual bodies in cybernetics, literature, and informatics*. Chicago: University of Chicago Press.

Heaven, D. (2013). Not like us: Artificial minds we can't understand. *New Scientist*, 219(2929), 32–35.

Jarrett, K. (2015). *Feminism, labour and digital media: The digital housewife*. London: Routledge.

Jin, D.Y. (2013). The construction of platform imperialism in the globalization era. *tripleC: Communication, Capitalism & Critique*, 11(1), 145–172.

Jones, S. (2014). People, things, memory and human-machine communication. *International Journal of Media & Cultural Politics*, 10(3), 245–258.

Kahn Jr, P. H., Ishiguro, H., Friedman, B., Kanda, T., Freier, N. G., Severson, R. L., & Miller, J. (2007). What is a human? Toward psychological benchmarks in the field of human-robot interaction. *Interaction Studies*, 8(3), 363–390.

Kerr, I. (2017). Speech before the House of Commons' Standing Committee on Access to Information, Privacy and Ethics. 4 April. https://techlaw.uottawa.ca/news/professor-ian-kerr-intervened-house-commons-discuss-pipeda

Klang, M. (2004). Spyware–the ethics of covert software. *Ethics and Information Technology*, 6(3), 193–202.

Lake, B. M., Ullman, T. D., Tenenbaum, J. B., & Gershman, S. J. (In press). Building machines that learn and think like people. *Behavioral and Brain Sciences*.

Le, Q.V., Ranzato, M., Monga, R., Devin, M., Chen, K., Corrado, G. S., Dean, J., & Ng, A.Y. (2012). Building high-level features using large scale unsupervised learning. *Proceedings of the 29th international conference on machine learning*, Edinburgh, Scotland.

Lessig, L. (2006). *Code: And other laws of cyberspace, version 2.0*. New York: Basic Books.

Leurs, K., & Shepherd, T. (2016). Datafication & discrimination. In M. T. Schäfer & K. van Es (eds.), *The datafied society: Studying culture through data* (pp. 211–231). Amsterdam: Amsterdam University Press.

McGeveran, W. (2009). Disclosure, endorsement, and identity in social marketing. *University of Illinois Law Review*, 2009(4), 1105–1166.

McMahan, H. B., Moore, E., Ramage, D., Hampson, S., & Aguera y Arcas, B. (2017). Communication-efficient learning of deep networks from decentralized data. *Proceedings of the 20th international conference on artificial intelligence and statistics (AISTATS)*, Fort Lauderdale, Florida.

Nissenbaum, H. (2004). Privacy as contextual integrity. *Washington Law Review*, 79, 101–139.

Pasquale, F. (2015). *The black box society: The secret algorithms that control money and information*. Cambridge, MA: Harvard University Press.

Pateman, C. (1988). *The sexual contract*. Stanford, CA: Stanford University Press.

Pateman, C. (2002). Self-ownership and property in the person: Democratization and a tale of two concepts. *The Journal of Political Philosophy*, 10(1), 20–53.

Pennachin, C., & Goertzel, B. (2007). Contemporary approaches to artificial general intelligence. In *Artificial general intelligence* (pp. 1–30). Berlin and Heidelberg: Springer.

Rawls, J. (2001). *Law of peoples*. Cambridge, MA: Harvard University Press.

Sample, I. (2017). Google's DeepMind makes AI program that can learn like a human. *The Guardian*, 14 March. www.theguardian.com/global/2017/mar/14/googles-deepmind-makes-ai-program-that-can-learn-like-a-human

Samuelson, P. (2000). Privacy as intellectual property? *Stanford Law Review*, 52(5), 1125–1173.

Shepherd, T. (2012). Persona rights for user-generated content: A normative framework for privacy and intellectual property regulation. *tripleC: Communication, Capitalism & Critique*, 10(1), 100–113.

Stabinsky, D., & Brush, S. B. (Eds.). (2007). *Valuing local knowledge: Indigenous people and intellectual property rights*. Washington, DC: Island Press.

Stone, P., Brooks, R., Brynjolfsson, E., Calo, R., Etzioni, O., Hager, G., … Teller, A. (2016). Artificial intelligence and life in 2030. One hundred year study on artificial intelligence: Report of the 2015–2016 Study Panel. Stanford University. https://ai100.stanford.edu/2016-report

Terranova, T. (2004). *Network culture: Politics for the information age*. London: Pluto Press.

Wachter, S., Mittelstadt, B., & Floridi, L. (2017). Why a right to explanation of automated decision-making does not exist in the General Data Protection Regulation. *International Data Privacy Law*, 7(2), 76–99.

Weinberg, L. (2017). Rethinking privacy: A feminist approach to privacy rights after Snowden. *Westminster Papers in Communication and Culture*, 12(3), 5–20.

Weston, J., Chopra, S., & Bordes, A. (2014). Memory networks. *Proceedings of the 3rd international conference on learning representations*, 7–9 May, San Diego.

Winner, L. (1977). *Autonomous technology*. Cambridge, MA: MIT Press.

Zuboff, S. (2015). Big other: Surveillance capitalism and the prospects of an information civilization. *Journal of Information Technology*, 30(1), 75–89.

Zuckerberg, M. (2017). Building global community. Facebook, 16 February. www.facebook.com/notes/mark-zuckerberg/building-global-community/10103508221158471/

14

UNTITLED, NO. 1
(HUMAN AUGMENTICS)

Steve Jones

In a 2011 paper, Kenyon and Leigh described the combination of transhumanism, captology, and Quantified Self (QS) efforts as "Human Augmentics," as, in their words, "technologies for expanding the capabilities and characteristics of humans ... [that could be] the driving force in the nonbiological evolution of humans" (p. 6758). For Kenyon and Leigh, Human Augmentics was centered on the principles of understanding human sensory, cognitive, and physical limits, and adopting advanced technology (hardware and software as well as shared intelligence and social networking) to develop technologies that may exceed those limits. The progenitor for their ideas was experience working with the medical rehabilitation community to use technology to encourage efforts by patients to use exercise to recover capabilities lost from, for instance, stroke or other brain injury. They closed their paper by writing that "Human Augmentics is a call to arms for the rehabilitation community to think outside of their boundaries—to think of the problem in terms of a larger interconnected ecosystem of augmented humans rather than a patchwork of disconnected sub-systems" (p. 6761).

In collaboration with a researcher from Rush University Medical Center, they and I began research using the principles of Human Augmentics to motivate inner-city African-American adolescents to use a daily inhaled corticosteroid to alleviate asthma exacerbations (Grossman et al., 2017). We began teaching a seminar in Human Augmentics at the University of Illinois at Chicago during the 2012 spring semester and have been teaching it annually since. Various Human Augmentics projects have emerged from the seminar, such as SpiderSense (Mateevitsi et al., 2013) and Audio Dilation (Novak et al., 2015).

Our teaching in Human Augmentics (as well as our research) was inspired from the start by the multidisciplinary nature of research that has been ongoing at the Electronic Visualization Laboratory (EVL) at the University of Illinois at

Chicago for over 40 years, and it incorporated visualization elements through-out. It is listed as a course in both Computer Science and Communication that is open to both graduate and undergraduate students. Enrollment has come from students in those two programs as well as Psychology, Bioengineering, Art and Design, Public Health, and other departments across campus. In addition to exploring the history and ethics of transhumanism, cyborgs, posthumanism, and other related topics, students form interdisciplinary teams to work on pro-jects of their own invention that employ the principles of Human Augmentics. They bring with them expertise in computer programming, rapid prototyping, design, human-computer interaction, persuasion, information studies, commu-nication, health, and medicine to focus on developing an idea into a prototype that demonstrates the capabilities and utility of their concept. At the end of the semester the students present the prototypes of their work and are critiqued by fellow students as well as by faculty from some of the departments represented in the seminar.

The experience of teaching Human Augmentics has led to a number of observations. First, during the initial couple of years of teaching, the seminar student projects largely hewed toward the health related. They tended to mir-ror then-current trends to create wearable fitness tracking devices, though, to their credit, the student projects sought to both provide more accurate data and to present the data in ways that could be more easily understood by users. The focus was typically on collecting data, visualizing it, then considering meth-ods to present it in a manner that might persuade the user to act in ways that could lead to healthier outcomes. While these were valuable efforts (and some continue to evolve) the most interesting projects, though they may have had an orientation toward health, began by imagining how technology interacts and interfaces with humans and then asking provocative questions concerning the human perception of the world outside and inside the body. It was those types of projects, typified by the aforementioned SpiderSense, for example, that caused us to reflect in a 2016 paper that,

> Human Augmentics is thereby about bending instead of blending. While the vision of ubiquitous and mobile computing that is seamless may invoke notions of cyborgification, the idea is not that technologies will merge with humans, subsuming them with, or into, the machine, rather that by altering forms of communication, devices more properly bend to the will of human users, reinforcing and expanding potential agency. The point of Human Augmentics is to develop communication between the human, machine, and environment premised on collaboration rather than co-option, engagement rather than estrangement, to increase human agency and a human's sense of agency, not to eradicate the human in pursuit of becoming something other. Human Augmentics, then, is focused on the intersections between human and machine, about the information that is

generated between agents, and the affordances that information provides to potentially increase agency.

(Novak et al., 2016)

The challenge for Human Augmentics, whatever technologies it may involve, is in iteratively assessing consequences for human agency in the design phase as well as during testing and implementation. In most instantiations of Human Augmentics, technology is presumed, is imagined, to operate as a check against some desired state of being, as a measure of a human function or activity, whether at an internal, cellular level (e.g. heart rate) or an external one (e.g. number of steps walked in a day), or as a "nudge" (Thaler & Sunstein, 2009) toward some future desired state. What cannot be overlooked is the notion of cognition, of the interpretation and understanding of the world inside and outside the self, and the mediation of the self in relation to the world.

In an essay on extended cognition, Clark and Chalmers asked, "Where does the mind stop and the rest of the world begin?" (1998, p. 7). This is an important question for those designing and studying Human Augmentics, as well, now, for anyone interested in artificial intelligence broadly, as AI increasingly mediates, appears as, and in some cases stands in for, human interaction and knowledge of, and interaction, with the world. The borders between human and machine are increasingly blurred; indeed, machines increasingly know more about us than do our friends and neighbors (Jones, 2014). Machines sense us, listen to us, observe us, and ceaselessly gather data about us, drawing inferences from patterns likely inscrutable to ourselves. Put simply, Facebook probably knows more about our neighbors than we do. It may know more about us than we imagine; that is, literally, it may put together pieces of data that are invisible to us, or, if visible, seemingly unconnected. And, in their connection, whether sensible or not, there are consequences.

Machines thus shape us and shape our world, while opportunities for us to shape them are few and far between. Stories abound of AI gone awry when learning to communicate with humans (e.g. Microsoft's Tay Twitter bot) and research has shown that machine learning contains and reinforces human biases (Caliskan et al., 2017). Human Augmentics portends an even closer relation to the machine, a cyborg manifestation, to riff on Haraway's seminal essay (Haraway, 1991). In relation to Human Augmentics, it is necessary to ask whether AI will not only contain and reinforce biases toward race, gender, and so on, but also make determinations, based on its observations and interactions with us, about our own biases and in turn attempt to alter (whether to reinforce or dispel) such biases. Put another way, will AI believe it knows our mind better than we do and act on that belief? Or, put still another way, what will make us aware of when and how we are being "nudged"?

For Clark and Chalmers the cognitive incorporation of the environment in human awareness results in a "coupled system" in which humans and external entities operate as a "coupled process [that] counts equally well as a cognitive process, whether or not it is wholly in the head" (p. 9). Another way to put this vis-à-vis Human Augmentics is that the processes that technologies engage are in the realm of thought and not only action. In other words, the notion espoused by Thaler and Sunstein (2009) of a "nudge" toward a desirable action or outcome, or the notion of "captology" espoused by Fogg (2003), is part of thought, not action. (These are also rather privileged and paternalistic notions, but that critique is more productively left for another day.) The augmentation is not simply or merely a physical one, whether it is a prosthesis, wearable, augmented reality headset, or some other technological component; it is an augmentation of the mind.

For those interested in Human Augmentics, this is a critical point to foreground. Clark and Chalmers envisioned coupled systems of human and information, and possibly human and human, but did not envisage human and machine. The human–machine coupling is a formidable system that ought to be interrogated in relation to the extended mind hypothesis they posited, particularly because this coupling carries with it agentic potentials they had not envisioned.

The balance between human agency and machine agency must therefore be a critical element in the design and evaluation of Human Augmentics technologies. As noted in the previous extended quote (Novak et al., 2016), the point of Human Augmentics is "to increase human agency and a human's sense of agency." Furthermore, the addition of "sense of agency" is meant not in the pejorative, as an "illusion" of agency, but rather to highlight the cognitive dimensions at the core of Human Augmentics. The systemic coupling of technology and mind, whether in a physical manner as may some day happen with brain implants that are routinely mooted as the new "five-years-out" technology, or in more metaphorical, yet no less real, ways, as when pundits (and parents) talk about smartphones being "welded" to the skin of their children, is a re-bordering of an already increasingly indistinct line not only between the self and the world but also between perception (*Noesis*, in Greek) and what is thought about (*Noema*). It is a phenomenological break with and through the machine. This re-bordering is occurring not simply because our attention is claimed by media and mediation (as Lippmann famously discussed in 1922 in relation to stereotypes and public opinion) but also because the self is a fully technically interpellated object, "not a separation of mind and body but an extraction of them" (Jones, 2014, p. 253).

Clark and Chalmers believe their notion of an "extended mind" will have consequences "in the moral and social domains" (1998, p. 18); likewise Human Augmentics has consequences in those domains, too, and it should be a goal of theorists and practitioners working in it to assay those consequences in their fullness. We live in an era in which we are increasingly networked, connected, to others, through use of "social media," yet we are also increasingly structured by those media into connections computationally, algorithmically, determined for us, and simultaneously disconnected from others. Indeed, the very term "social media" also implies and incorporates its opposite, and it is a term whose history, evolution, demands to be unpacked and examined. What opportunities are offered for agency are generally binary: opting in/out, using/refusing. A significant concern for Human Augmentics is whether it will recover agency for humans or will continue on the course already set by social media. If we recognize that what distinguishes Human Augmentics from other technologies is the awareness of the systemic coupling of technology and mind and the presentation of that coupling as an informative, interpretable, actionable potential, we may yet find a way to articulate the self and the world, the digital and the analog, in ways that we value.

References

Caliskan, A., Bryson, J. J., & Narayanan, A. (2017). Semantics derived from language corpora contain human-like biases. *Science*, 356, 183–186.

Clark, A., & Chalmers, D. (1998). The extended mind. *Analysis*, 58(1), 7–19.

Fogg, B. J. (2003). *Persuasive technology: Using computers to change what we think and do.* San Francisco, CA: Morgan Kaufmann.

Grossman, B., Conner, S., Mosnaim, G., Albers, J., Leigh, J., & Kenyon, R. (2017). Application of human augmentics: A persuasive asthma inhaler. *Journal of Biomedical Informatics*, 67, 51–58.

Haraway, D. (1991). *Simians, cyborgs and women: The reinvention of nature.* New York: Routledge.

Jones, S. (2014). People, things, memory and human–machine communication. *International Journal of Media & Cultural Politics*, 10(3), 245–258.

Kenyon, R., & Leigh, J. (2011). Human augmentics: Augmenting human evolution. *Conference Proceedings of the IEEE Engineering Medicine Biology Society* (pp. 6758–6761). Boston, MA, August 30. doi: 10.1109/IEMBS.2011.6091667

Lippmann, W. (1922). *Public opinion.* New York: MacMillan.

Mateevitsi, V., Haggadone, B., Leigh, J., Kunzer, B., & Kenyon, R. V. (2013). Sensing the environment through SpiderSense. *Proceedings of the 4th augmented human international conference* (pp. 51–57). AH 13. New York: ACM. doi: 10.1145/2459236.2459246

Novak, J. S., Archer, J., Shafiro, V., & Kenyon, R. V. (2015). Audio dilation in real time speech communication. *The Journal of the Acoustical Society of America*, 137(4), 2303. doi: 10.1121/1.4920407

Novak, J. S., Archer, J., Mateevitsi, V., & Jones, S. (2016). Communication, machines & human augmentics. *communication* +1, 5. http://scholarworks.umass.edu/cpo/vol5/iss1/8. doi: 10.7275/R5QR4V2D.

Thaler, R., & Sunstein, C. (2009). *Nudge: Improving decision about health, wealth and happiness.* New York: Penguin Books.

INDEX

Note: index entries in the form 134n2 refer to page and endnote numbers.

For Product Safety Concerns and Information please contact our EU
representative GPSR@taylorandfrancis.com
Taylor & Francis Verlag GmbH, Kaufingerstraße 24, 80331 München, Germany